G.I.BUDKER

Reflections & Remembrances

G.I.BUDKER

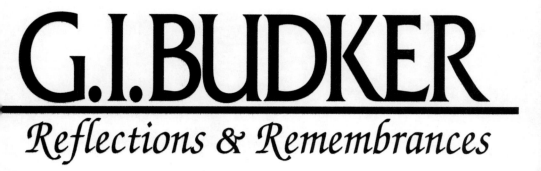

Reflections & Remembrances

EDITED BY

BORIS N. BREIZMAN

·

JAMES W. VAN DAM

AIP
PRESS

American Institute of Physics   New York

American Institute of Physics
500 Sunnyside Boulevard
Woodbury, NY 11797-2999

Library of Congress Cataloging-in-Publication Data

Akademik G.I. Budker. English.
    G.I. Budker : reflections and remembrances / edited by Boris N.
Breizman and James W. Van Dam ; translated by Natasha Breizman.
        p. cm.
    Translation of: Akademik G.I. Budker.
    Includes bibliographical references and index.
    ISBN 1-56396-070-2
    1. Budker, G. I. (Gersh Itskovich), 1918-1977. 2. Nuclear
physics--Research--Soviet Union--History. 3. Nuclear physicists-
-Soviet Union--Biography. I. Breizman, Boris N. II. Van Dam,
James Walter, 1948- . III. Title.
QC774.83.A4213 1993
539.7'092--dc20
[B]   93-22850
                                                              CIP

Contents

Foreword

This volume is a translation of a memorial volume published in 1988 in memory of Academician Gersh Itskovich Budker, who died July 4, 1977. The original was edited by Academician A.N. Skrinsky, who worked with Budker through much of his productive life. Publication of this book is extremely timely in view of the great intensification of cooperation between Western and Russian scientists at the end of the Cold War. It is important for American physicists to become acquainted with one of the true greats of Russian science.

Budker was a physicist who was often ahead of his time both scientifically and politically. As is documented in this volume and in Budker's extensive writings, he initiated many devices which only came to full fruition when the original ideas were elaborated by others at a much later time. He perceived of science as a truly international enterprise, only to be frustrated by the political barriers of his time. I hope that this book will serve to make knowledge of Budker's contributions available to a wider audience who can benefit from both the substance of his innovative ideas and the knowledge about a truly great man.

Wolfgang K.H. Panofsky
Stanford Linear Accelerator Center
Stanford University

Preface to the English Edition

This is a book about a scientist, but it is not a scientific book. No book can present a scientist better than his own papers can, as Albert Einstein once said. Rather, this is a book about a person — an incredibly talented person and marvelous raconteur and wise teacher and, last but not least, a physicist.

Indeed, G.I. Budker was one of the premier figures of Soviet physics. Among his numerous scientific contributions were the stabilized electron beam, the magnetic-mirror plasma trap, colliding beams, and electron cooling. His genius extended to several areas: high-energy physics of elementary particles, nuclear physics, accelerator technology, and plasma physics and controlled fusion. Because of his unconventional talent for inventing practical applications of relativistic physics, Lev Landau cleverly complimented Budker with the nickname "engineer of relativity." In fact he found such wide-ranging commercial uses for particle accelerators in industry, agriculture, and medicine that he was able to subsidize much basic research.

What makes Budker interesting to a wider audience, however, are his irreverent, yet sage, sense of humor and also the unorthodox, yet effective, views he developed on how doing science should be organized. When he founded and built up the world-famous Institute of Nuclear Physics in Siberia (far from the madding crowd of Moscow bureaucrats), he instituted a round table system of consensus management, based on ability rather than seniority. His ideas for rearing the next generation of young scientists were similarly innovative, and their proof of principle has been evidenced at his Institute. And, through it all, his colorful eccentricities, quotable witticisms, and nonstandard insights made him an unforgettable character, a much revered director, and a beloved mentor to many.

The reminiscences in this book vividly describe all of these multiple facets of Budker's talents and personality. As the preface to the 1988 Russian-language edition states, the remembrances of Budker written by his domestic and international colleagues are largely arranged in chronological order. In the present English edition, this arrangement has been made explicit by their grouping according to the two periods of, first, Budker's early work in Moscow at the Kurchatov Institute of Atomic Energy (after World War II until 1957) and, second, his work as scientist, director, and entrepreneur at the Institute of Nuclear Physics in Novosibirsk (from 1958 until his death in 1977).

Several additional articles and materials are included in this English edition. We are very thankful to Roald Sagdeev, Norman Rostoker, and Frederick Mills, longtime acquaintances of Andrei Budker, for contributing new articles, and to Wolfgang Panofsky for writing the foreword. From a published collection of papers presented at a special memorial session of the American Physical Society in Washington, D.C., on April 25, 1978, and those presented on the same day at a memorial symposium in Novosibirsk to mark Budker's sixtieth birthday[1] (and the twentieth anniversary of the founding of the Institute of Nuclear Physics), we have excerpted portions that deal with personal remembrances about Budker. Also, to the original six selected popular essays written by Budker himself, we have added the record of his concluding remarks at the 1968 IAEA Conference, in which he

---

[1] Also on the occasion of his sixtieth birthday, an issue of *Soviet Journal of Plasma Physics*, Vol. 4, No. 3, May-June 1978 (Russian original: *Fizika Plazmy*, Vol. 4, No. 2, May-June 1978) was specially dedicated to his memory.

issued a clarion call to "get going" with fusion reactor studies. Finally, we have included a listing of Budker's publications. Some explanatory footnotes throughout the book were added in the translation process, supplementing the original footnotes of the authors and those of the Russian editors.

The editors express their gratitude to Natasha Breizman for her fine work in translating the materials from Russian into English, and to Suzy Mitchell for her word-processing expertise in preparing the camera-ready copy. Most of the translated articles were kindly checked and, in some cases, supplemented by the authors. Both of our home institutions have been very generous in providing support for this project, and we wish to thank many colleagues at both places and elsewhere. In particular, we appreciate the willingness of the members of the editorial committee for the original Russian book and its editor-in-chief, Alexander Skrinsky, present director of the Budker Institute, to make the publication of this English version possible. With composing Budker's bibliography, we especially appreciate the efforts of Mrs. L.P. Zhuravleva, librarian at the Budker Institute. We thank Andrei Kudryavtsev and Simon Eidelman for various general assistance, and Buff Miner for help with software. We also gratefully acknowledge the critical assistance of John Zumerchik, book editor, Andrew Prince, book coordinator, and Maria Taylor, publisher, of AIP Press, of the American Institute of Physics.

We are pleased to make available in English this book about a remarkable person — scientist, organizer, director, mentor, and sage — Academician G.I. Budker.

Boris N. Breizman
Budker Institute of Nuclear Physics
Russian Academy of Sciences

James W. Van Dam
Institute for Fusion Studies
The University of Texas at Austin

Preface to the Russian Edition (1988)

This book is a collection of remembrances of the eminent Soviet physicist and academician, Gersh Itskovich (Andrei Mikhailovich) Budker, written by his fellow scientists, comrades, and students, who all knew him through joint work with him, and by foreign colleagues.

Andrei Mikhailovich lived a relatively short life in the scientific world, not much longer than 30 years, but it was a very colorful one. However, his well-known contributions during this time constitute many glorious pages in the history of Soviet physics.

The pediment of the main building of the Institute of Nuclear Physics in Novosibirsk is decorated by a strip of lights, usually turned on during holidays. Then, along the top of the building, two shining lights rush to meet each other. Having met in the middle of the sign, they burst into a bright star. This is a symbolic representation of colliding beams and, at the same time, the emblem of the Institute of Nuclear Physics (INP) of the Siberian Division of the Academy of Sciences of the U.S.S.R., which gave to world science the most effective modern method for the study of the structure of the microscopic world. At that time, the end of the 1950's, the boldness of Andrei Mikhailovich was necessary, his unwavering belief in success, with which he infected his colleagues in order to take up the realization of what seemed to be at that time the absolutely hopeless idea of electron-electron colliding beams. And thus it was for all those years. He propounded what "sober-thinking" skeptics — "realists" — quickly declared to be fantastic, the fruits of the imagination of a theorist, far from the technology of the physical experiment. However surprising this sounds today, Andrei Mikhailovich really was a theoretical physicist, although the overwhelming majority of his work was related to experimental physics, where he made a large number of original proposals and inventions that have enriched the technology of physical experiments.

Probably the most striking example of Budker's "crazy" ideas is the electron cooling method, which allows one to obtain dense ("cold") beams of heavy charged particles such as protons, antiprotons, and ions. Until this method was demonstrated at the INP, not a single laboratory in the world risked an attempt at it, although it was very promising. It seemed to be a hopelessly complicated problem.

The scientific interests of Andrei Mikhailovich lay in two areas: high-energy physics and the physics of controlled thermonuclear fusion. At first glance the two areas seem very far removed from each other. But the talent of such a scientist and his gift of foresight were needed to make clear in the course of time how much these two branches of physics have in common, both dealing primarily with charged particles. It is curious that nowadays even in the scientific nomenclature, the specialty "Physics of charged particle beams and accelerator technology" has already appeared, which is given to scientists working in both of these fields.

Today the concept of "scientist-organizer" has become quite common. Sometimes this title is applied to people who are not very strong researchers, who have not made any outstanding discoveries, but who are able to organize others. For Andrei Mikhailovich, they came together — both the role of scientist and the role of organizer — and were inseparable. The entire structure of the Institute of Nuclear Physics and the formation of a distinctive style of work and of interrelations within the collective were the result of his original ideas in the area of organizing scientific investigations. The INP is often called "The Institute of the Round Table" because of this method for constant, active interaction among scientific colleagues of all ranks, which was thought up and brought about by Andrei Mikhailovich. And so it was for all areas, from experimental production and the supplies department to the scientific laboratories and design department. Budker's ideas are recognizable even in the organizational structure of the Institute: experimental laboratories whose subjects are similar are put not in separate divisions, but in a united laboratory. "Division comes from the word divide, whereas we should be united in our work," said Andrei Mikhailovich more than once.

Above all there was meticulous, everyday work with co-workers, the solution of all possible problems, and conflicts, but only directly related to the work. Invariably there occurred precise, strongly individualistic selection of the persons assigned for carrying out tasks and creative initiatives in all areas with everyone. Andrei Mikhailovich liked to repeat, "In our work, even the lathe operator turning a bolt should understand that he is not simply turning a bolt, but making an Accelerator."

This collection opens with a series of popular scientific articles and essays by A.M. Budker. It is necessary to mention in particular his essay in *The Age of Knowledge*, an ardent appeal to young people, with his lifelong credo "Beware of calms!" which resounds especially strongly today.

The circle of the authors of the remembrances is extremely wide. People of different characters and temperaments, of different and at times opposite points of view, were unified in the wish to preserve in memory the dear features of their teacher, friend, and colleague. The editors tried their best to preserve the originality of each author's style in the preparation of the materials for the collection. The arrangement of the articles is determined by the chronology of the remembrances. The beautiful photographs were donated by R.I. Akhmerov, V.N. Bayev, V.N. Davidenko, A.I. Zubtsov, G.D. Kustov, V.T. Novikov, V.V. Petrov, A.N. Polyakov, A.P. Usov, and A.I. Shlyakhov. Most of the photographs are published here for the first time. Unfortunately, with some of the photographs, we did not succeed in finding out who took them.

The editors express their deep gratitude to all who participated in the preparation of this book.

Chief Editor: A.N. Skrinsky

Editorial Committee: G.I. Dimov, E.P. Kruglyakov, D.D. Ryutov,
                       V.A. Sidorov, B.V. Chirikov, and I.N. Meshkov

In Memoriam

In Memory of Academician Budker

A.P. Alexandrov, L.M. Barkov, S.T. Belyaev, Ya.B. Zel'dovich, B.B. Kadomtsev, A.A. Logunov, M.A. Markov, D.D. Ryutov, V.A. Sidorov, A.N. Skrinsky, B.V. Chirikov

May 1, 1978 marks the sixtieth anniversary of the birthday of the eminent Soviet physicist, academician Gersh Itskovich (Andrei Mikhailovich) Budker, founder and director of the Novosibirsk Institute of Nuclear Physics, recipient of the Lenin Prize and the U.S.S.R. State Prize.

Andrei Mikhailovich was born on May 1, 1918 in the village of Murafa in the Shargorodski District of Vinnitsa Region into the family of a rural laborer. In 1936, having finished secondary school in Vinnitsa, he entered the physics department of Moscow State University.

This obituary was originally published in *Uspekhi Fizicheskikh Nauk*, Vol. 124, No. 4, pp. 731-735, April, 1978. [An English translation appeared in *Soviet Physics—Uspekhi*, Vol. 21, No. 4, pp. 369-372, April, 1978.]

The new physics with its extraordinary theories captivated Andrei Mikhailovich from the very beginning. He took to it right away, without looking back to "common sense" and classical notions. Subsequently he always related negatively to attempts to return to "the good old days" of classical physics. Instead, he managed to develop his own imagination so that the theories of relativity and quantum mechanics, which he understood keenly and deeply, became for him not just understandable but also natural and visualized, theories with which he was able to "work." Not without reason one of the divisions of the special curriculum that Andrei Mikhailovich taught in his last years at Novosibirsk University was called "Relativistic Constructs."

Andrei Mikhailovich did his first scientific work as a student under the guidance of I.E. Tamm. It was dedicated to the problem of finding the energy-momentum tensor of an electromagnetic field in moving media. It is possible that already in this problem Andrei Mikhailovich felt the enormous hidden difficulty and, together with that, the beauty and inexhaustible possibilities of complex systems, which afterwards he researched and mastered so skillfully in his works on high-current accelerators and thermonuclear reactors.

Andrei Mikhailovich graduated from the university in 1941 and immediately after his last degree examination went into active duty in the army. In an anti-aircraft field unit he made his first invention, improving a system for guiding anti-aircraft fire. The commander of the unit named the instrument created by Budker the AMB.

After the end of the Great Patriotic War, Andrei Mikhailovich entered the theoretical department of Laboratory Number 2, the famous "Two," headed by I.V. Kurchatov (today it is the I.V. Kurchatov Institute of Atomic Energy). Still a very young physicist, he took an active role in solving the atomic energy problem. Under the leadership of I.V. Kurchatov and A.B. Migdal he completed a series of works on the theory of the finite uranium-graphite lattice, as well as on the kinetics and control of atomic reactors.

In connection with the construction of a record large (for that time) proton accelerator in Big Volga (now the city of Dubna), the focus of Andrei Mikhailovich's interests shifted to the theory of cyclical accelerators. He was the first who paid attention to resonance processes in accelerators and studied them in detail, who created techniques for calculating the shimming of a magnetic field, and who proposed original ways for efficiently extracting beams from an accelerator. These works were awarded the U.S.S.R. State Prize in 1951. However, Andrei Mikhailovich himself already clearly understood that further development of accelerator technology was impossible without a consideration of the collective processes in the accelerated beam of particles. Together with his first students he began actively to develop a theory of such processes, in essence laying the foundation for a new field of physics: the physics of a relativistic plasma. In particular, he created the theory for the relativistic kinetic equation and found its solutions in the so-called "anti-diffusion" approximation (for infrequent collisions).

Andrei Mikhailovich was never a "pure" theorist. Besides the inclinations of his own character, another reason for this was the enormous influence of the Kurchatov school. Therefore, he immediately tried to apply the remarkable properties of relativistic plasma to solve the pressing problems of accelerator technology. Moreover, and this is, no doubt, one of the most remarkable features of Andrei Mikhailovich's creative personality, the gradual improving of existing accelerators never satisfied him. He persistently searched for fundamentally new approaches and solutions in this field. And he succeeded in theoretically discovering an incredibly beautiful structure, formed from relativistic electrons and ions, which he called the stabilized electron beam. After the report about this work at the Geneva Conference in 1956, the name of A.M. Budker became widely known, and his ideas caused great interest among physicists of many countries.

Around this time Andrei Mikhailovich proposed an original approach to the solution of another urgent problem of physics: the problem of controlled thermonuclear fusion. His approach was based on the use of a plasma trap with "magnetic mirrors" ("plugs") and laid the beginning of all so-called "open" fusion systems.

Andrei Mikhailovich was extremely impatient to take up the implementation of all these ideas immediately. However, these ideas were too complicated, almost fantastic, whereas Andrei Mikhailovich himself was only a theoretician. And then he took probably the most important step in his life, a very brave and extraordinary step — it is better to say "leap" into the unknown instead of "step": he decided to head up a group of enthusiasts — experimentalists and engineers — who were ready to implement his ideas. Andrei Mikhailovich took this step not without inner hesitation and even fear, but he nevertheless made the decision, in spite of persistent advice and admonitions from many close friends. Although having no experience at all in the organization of experimental research, but unconstrained by tradition, Andrei Mikhailovich also suggested his own original ideas in this area: how a creative scientific group should live and develop. Thus was born the Budker school. In the beginning, in 1953, it was a small group of only eight people. But results were soon forthcoming — after only a few years a betatron accelerator with a current of up to 100 amperes was constructed, which exceeded by two orders of magnitude the currents of the best accelerators of that time. Andrei Mikhailovich's small group grew into one of the largest laboratories in the Institute of Atomic Energy, the Laboratory of New Acceleration Methods, and in 1958 it was transformed into the independent Institute of Nuclear Physics of the young Siberian Division of the U.S.S.R. Academy of Sciences.

Nonetheless, they did not succeed in creating a stabilized beam — the technical difficulties proved to be insurmountable; this problem is still awaiting its future resolution. Probably Andrei Mikhailovich understood this before anyone else. What was to be done? The group was already fairly large and working intensively with full efficiency. Which way to go? Where should this stream of creative energy be directed? And he found the solution — colliding beams! The idea of the collision of two accelerated beams had already been mentioned in the literature, and the enormous energy advantages of colliding beams in the creation of new heavy particles were clear. However, most people regarded this idea as a joke or an unattainable dream. Actually, the role of the dense target of a usual accelerator is played here by a rarified colliding beam, whose density was several orders less than the density of the highest vacuum (at that time). However, the experience accumulated in Andrei Mikhailovich's laboratory concerning the study of the physics and creation of the new technology of intense relativistic beams also opened the way for the solution of this fantastic problem, yielding colliding beams of electrons, and then of electrons and positrons. Of course, it was very risky to attempt such work, but it was a justified risk, without which there would have been no notable accomplishments. The decision for the creation of devices with colliding beams was not made immediately. I.V. Kurchatov provided great support for it, believing in Andrei Mikhailovich's bold ideas and in the creative power of his group. Thus arose the basic direction of research of the Institute of Nuclear Physics and a new area in the experimental physics of elementary particles. Andrei Mikhailovich was one of the pioneers of this field in international physics.

The first device with colliding electron beams (VEP-1) was completed in Novosibirsk. In 1965 the first experiments on the verification of quantum electrodynamics to distances on the order of 10^{-13} cm were carried out on it. Meanwhile, Andrei Mikhailovich suggested the new, even more exciting idea of creating a device with colliding electron-positron beams. The central problem here was the storage of a significant (tens of milliamperes) positron current, which required the assured "production" of positrons in large numbers. Andrei Mikhailovich persistently searched for a solution to this problem, going over dozens of different options, inventing, analyzing, improving. And thus was born the simple scheme for multiple storage of positrons on a magnetic track with the use of radiative damping of beam oscillations as the result of synchrotron radiation. The decisive element in this plan was high aperture positron optics, using original parabolic lenses and providing the effective collection of positrons after a converter. Thus arose the VEPP-2 device, on which the first experiments in the world with colliding electron-positron beams were done in 1967. This direction turned out to be very fruitful, and today a significant part of all fundamental information on elementary particles has been gained precisely from these experiments. In particular, this method turned out to be very effective for the carrying out of "pure" experiments and for the study of strong interactions. In 1967 Andrei Mikhailovich and his colleagues were awarded the Lenin Prize for this work.

Work on colliding beams was first reported in 1963 at the International Conference on Accelerators in Dubna and caused great interest. Immediately after the conference, the first group of foreign scientists visited the Institute of Nuclear Physics. This signified the beginning of close and fruitful cooperation by the Institute with many scientific centers in Europe and America, cooperation which since then has continually broadened and deepened and to which Andrei Mikhailovich always attached great significance.

In 1974 experiments began on a new device, VEPP-2M, which in the old energy domain (up to 2×700 MeV) possesses very high "luminosity" and consequently a high frequency of collisions between electrons and positrons. The luminosity of VEPP-2M ($2 \times 10^{30}\,\mathrm{cm}^{-2}\,\mathrm{sec}^{-1}$) to date exceeds the luminosity of all other devices in this energy domain by more than an order of magnitude. This result was achieved by the formation of incredibly narrow beams: at the point of collision their width was only 10 microns!

In 1966 Andrei Mikhailovich proposed an effective method of damping incoherent fluctuations in beams of heavy particles, for which radiation damping is practically nonexistent. The idea behind the method is quite simple: parallel to a beam of heavy particles, a beam of electrons moves with the same average velocity and sufficiently low temperature. Then the frequency of binary collisions sharply increases, and the heavy particles "cool down," giving energy to the electrons. This method was named electron cooling. The effectiveness of this method was demonstrated on the experimental device NAP by cooling a proton beam with an energy of about 100 MeV: in one-tenth of a second under these conditions, the proton successfully cooled down to a temperature of $1/20\,\mathrm{eV}$.

Many physicists eagerly awaited this result. After all, the method of electron cooling allows compression of a beam of heavy particles in the transverse direction, and consequently allows multiple storage of such particles on a magnetic track, which opens up the possibility for the creation of a device with colliding proton-antiproton beams. The news of the successful realization of electron cooling quickly spread among physicists of many countries. Several scientific centers entered into cooperation with the INP to master this method.

Examining different versions of proton accelerators for future proton-electron and proton-antiproton colliding beams, Andrei Mikhailovich proposed the new method of charge-exchange injection. The idea for the method consists of injecting into the accelerator negative ions of hydrogen, which then, losing electrons, become protons and thereby are irreversibly captured onto the magnetic track. Experiments done at the INP confirmed the high efficiency of this method. Andrei Mikhailovich suggested as well to compensate the beam of protons, circulating in the accelerator, with electrons in order to exceed the proton space-charge limit. Experiments showed that, under certain conditions, in particular with a sufficiently dense plasma inside the beam, the latter will remain stable. By this method a current was stored that exceeded by an order of magnitude the space-charge limit for an uncompensated proton beam.

Already, when creating the first devices with colliding beams, Andrei Mikhailovich suggested using the unique properties of the synchrotron radiation of such beams for carrying out a wide class of experiments in the areas of chemistry and biology. Nowadays the Center for Synchrotron Radiation, where scientists work from many organizations in different cities in the Soviet Union, is operating at the INP. On the VEPP-2M and VEPP-3 devices, special channels for synchrotron radiation were built, fitted with unique detection instruments, also created at the INP. A characteristic example of such research are experiments conducted jointly with the Institute of Biological Physics of the U.S.S.R. Academy of Sciences on the study of the dynamics of the structural rearrangement of molecules of living frog muscles in the process of contraction. In the course of one contraction cycle, lasting approximately 0.1 sec, it is possible to take 60 consecutive X-ray photographs.

Having advanced the idea of confinement of hot plasma in a trap with magnetic mirrors, Andrei Mikhailovich constantly returned to it, examining different aspects of "open" thermonuclear systems. After an initial period of disappointment, caused by the abundance of plasma instabilities, Andrei Mikhailovich was one of the first to concentrate his efforts in this area, with a deeper and more serious study of plasma physics. In particular, he suggested investigating the behavior of a thermal plasma that is from the beginning in thermodynamic equilibrium, in order to avoid the turbulence that is characteristic of plasma heating by high-power electrical discharges.

Ten years passed of intensive research in plasma physics, carried out in many laboratories all over the world. Andrei Mikhailovich then came to the conclusion that a new phase had arrived in solving the thermonuclear problem. In 1968 at the Third International Conference on Plasma Physics and Controlled Thermonuclear Fusion, which took place in Novosibirsk, he called on physicists to get down directly to the business of developing a thermonuclear reactor. His idea was that plasma physics had already been studied well enough to make it possible to search for the solution for the first physics fusion reactor. This call exerted great influence on the development of thermonuclear research and, in particular, was the beginning of serious study of the engineering problems of future thermonuclear reactors.

Andrei Mikhailovich himself suggested a new approach toward the solution of this problem: the essence of this approach consisted of using a magnetic field only to decrease the transverse thermal conductivity of the plasma, while its pressure is confined by ordinary walls. For the reduction of the heat conductivity along the field it was suggested to use a novel "multiple mirror" magnetic field configuration. The idea was that the speed of the plasma as it expands in the longitudinal direction will sharply decrease if the mean free path length of the particles becomes of the order of the distance between neighboring mirrors. Experiments done at the INP confirmed the effectiveness of this method for the thermal insulation of a plasma.

In accordance with the ideas of Andrei Mikhailovich, high-power generators of pulsed relativistic beams, the first to be used to heat plasma, were created at the INP. The use of ultra-pure water as a dielectric in the energy storage banks for the generators of such beams played a significant role in this. This work exerted an important influence on the development of the technology of ultra-high-power energy sources.

All these diverse activities should have, it seems, completely absorbed Andrei Mikhailovich. But for him it was not enough — this is, no doubt, one of the brightest features of his character — he persistently searched for immediate applications of all that his institute knew and could do, to current urgent problems of the national economy, and he found it — industrial accelerators! These modest devices do not stagger the imagination by their size or the energy of their particles. However, they are very necessary for industry, and Andrei Mikhailovich gave a significant amount of his time, energy, and inventiveness to the development of this direction at the Institute. Beginning in 1963, under his direct leadership, a whole series of special electron accelerators was developed and produced, with an average power of several kilowatts up to a megawatt and electron energies of several hundred kilovolts up to 2 MeV for radiation processing of materials. This allowed moving to a fundamentally new production technology in widely different areas of the national economy. Several characteristic examples: a sharp increase in thermal stability of polyethylene insulation; the manufacture of special thermosetting hoses of polymers that "remember" their original size; disinsectization of grain; disinfection of sewage; cutting and welding of metals; and many others.

Thus the distinctive scientific themes and organizational structure of the Institute arose and are still developing. It is necessary to emphasize that the success and achievements of the Institute, widely known in our country and also far abroad, were the result not only of the fundamental ideas of Andrei Mikhailovich, but also of his daily, indefatigable work, his tireless searches for original solutions for a multitude of particular and, at first glance, small problems that are unavoidable in any major undertaking.

Andrei Mikhailovich believed that the best method for solving a difficult problem, whether it be in physics, technology, or organization, is a collective search by way of constant discussion of all possibilities from all angles, even the most fantastic. Such collective creative effort should, of course, be supplemented by intense individual work by everyone who takes part. Andrei Mikhailovich himself worked extremely intensively, everywhere and all the time, not knowing rest, even in the last years of his life when he was very ill. As a rule, he always found the necessary solution.

Andrei Mikhailovich was not only an outstanding physicist, but also a wonderful Teacher. The need to teach others his favorite science, discovering not only physical laws but also human talents and bringing up future researchers, was an integral feature of his versatile personality. Andrei Mikhailovich began teaching when he was still a very young physicist at the newly organized Physics and Technology Department of Moscow University. It was there that he selected his first students. Upon moving to Novosibirsk, Andrei Mikhailovich took an active part in the organization of Novosibirsk University. He set up an original course of general physics and organized and headed the Department of General Physics and then also the Department of Nuclear Physics. Under his initiative a special Physics and Technology Department was created at the Novosibirsk Electrotechnical Institute, which trained many talented engineering physicists. Andrei Mikhailovich and his students actively participated in the organization and carrying out of the All-Siberian Physics and Mathematics Olympiads for secondary school students, and taught at the Physics and Mathematics Preparatory School of Novosibirsk University.

Memorial plaque on the front of the main building of the Budker Institute of Nuclear Physics.

But, of course, the most important school of future researchers and engineers is the Institute of Nuclear Physics, where students take full part in scientific work, seminars, discussions, and arguments. And especially fortunate are those who were lucky enough to work directly with Andrei Mikhailovich. He recognized in science neither organization charts nor rank and demanded only one thing — do not "make the air tremble," but put a thought into every word; do not remain a captive of formal syllogisms. As a reward for this, he generously revealed his most treasured thoughts, worked out over many nights of intense concentration, original concepts, unexpected parallels, and analogies, the wise edifications of a man who lived a great and complex life. Especially interesting were the sessions of the Scientific Council of the Institute, which meets every Wednesday at 12 o'clock noon at the Round Table, symbolizing the intolerance of purely administrative decisions in science. Attempting to involve as many staff members as possible, including very youthful ones, in the discussion and deciding of the most important scientific and organizational questions concerning the work of the Institute, Andrei Mikhailovich created in his last years three more topical scientific councils, which also meet every week.

The Institute was the favorite "child" of Andrei Mikhailovich. He was never simply the director. The Institute was the embodiment of his creative design in physics, technology, and the organization of science. The Institute is also a new scientific school in high-energy physics, accelerator technology, and plasma physics, a school with its own traditions, principles, and ideals. But the Institute is also an enormous collective of scientific members and engineers, of workers and service employees, a collective with its own complicated life, which Andrei Mikhailovich understood so well and directed so skillfully. He combined in himself the scientist, the inventor, and the organizer. In this fruitful synthesis lies the foundation of the Institute and the hallmark of its success, both in the past and the future.

Selected Essays of G.I. Budker

Science Expands Horizons: Microworld Economics

G.I. Budker

I recall a conference that was held 20 years ago, after the first synchrotron had been set up in Dubna. The agenda was a final summing up of the results of this construction. From the conference hall, one could see a mysterious concrete structure — an enormous cube, as high as a multi-story building, but with no windows or doors. Inside of it was a machine that generated particles whose energy was hundreds of times more concentrated than the energy of nuclear reactions. Great hopes were placed on this first large accelerator.

"Why don't we remove the railroad track now?" suggested one of the participants at the conference.

A railroad branch track had been built for the purpose of the assembly of the device, and now there was practically no need for it anymore.

Originally published in *Pravda*, February 27, 1969.

"What do you mean, 'remove'?" indignantly exclaimed one of the construction people, his good feelings insulted. "How do you plan to transport the output?"

He did not understand very much about nuclear physics (which was quite a new subject at that time), but he had no doubt that such a complex facility, a marvel of construction technology, would produce something very useful by the ton. However, the 20-year "output" of the accelerator consists of photographs of nuclear reactions and also some curves and numbers that characterize features of elementary particles. All of it, in the end, can be put inside one briefcase. The accelerator "manufactures" ideas and knowledge — which, in our day, are the most valuable commodities.

Nevertheless, was the builder who asked this question about the real products of the accelerator entirely wrong?

It seems that mankind, having passed from the era of chemical energy into the era of nuclear energy, needed to stop and reflect. The leap was too great. In fact, however, nothing of the sort happened: man has been attracted by an even more powerful force.

Cosmic rays with an energy much higher than the energy of nuclear reactions arrive from the abyss of the Universe and shower upon the Earth. These rays give birth to new unknown particles in the atmosphere, from which physicists have gotten the idea that the atomic nucleus is not the terminal boundary of knowledge. However, man does not yet control the Universe, and cosmic rays do not fall upon the Earth at his discretion. Moreover, they are inconvenient for experiments. Therefore, physicists have created their own generators of cosmic rays. The first such accelerator to energies of more than a billion electron volts was even named the "Cosmotron."

Nowadays, accelerators with much higher energies are operating around the world: for example, the 30 billion electron volt accelerator at Brookhaven (in the U.S.A.), the 28 billion electron volt accelerator at Geneva, and finally, the largest of all, our own 70 billion electron volt accelerator at Serpukhov. Accelerators with super-high energies allow many discoveries to be made, thus bringing fundamentally new concepts and information into science. For example, thanks to them, antiprotons and antineutrons were discovered. Thereby was proved the existence of antimatter — an unusual type of matter — the fuel of the future, a billion times more efficient than modern rocket fuel.

The development of science requires the use of particles with even higher energies. Higher energies means bigger accelerators. But isn't the bigness we have already enough? Modern accelerators are several kilometers long and consist of magnets weighing many tons that are set on top of strong foundations and precisely positioned to within a few microns. These machines are so complex and expensive that constructing one often becomes a national task. The 200 billion electron volt accelerator now in the process of being constructed in the U.S. is, in fact, even called a "national" accelerator.

Still, a high-energy accelerator is, in the final analysis, nothing more than an instrument, the "microscope" of modern physics. There is something unnatural when building an instrument requires the efforts of an entire country. There is also the sad fact that only a small fraction of the particle energy, which has been obtained with such an effort, is actually utilized in accelerators. Imagine a locomotive colliding with a fly. The locomotive hardly gives up any of its energy at all in this collision. Similarly, a particle traveling at almost the speed of light, whose mass, according to the theory of relativity, increases many thousands of times, also gives up a tiny part of its energy when it interacts with a particle at rest. This inefficiency of collisions increases as the energy of an accelerator goes up, and eventually sets a limit to the capabilities of the usual acceleration methods.

The solution was found in the so-called method of colliding beams. Let us make two particles, moving in opposite directions, collide with each other. It does not matter that their masses increase with speed; they collide as equals, and all their energy is expended in the interaction, giving birth to new particles. This method has no waste — all the expended effort goes into work. Implementing this method, however, required almost ten years of strained labor, much patience, and much skill on the part of the experimentalists. Two years ago in Novosibirsk, the first experiments were carried out on a device with colliding electron and positron beams. A year later, a similar device began operation in France, and more recently another in Italy.

The most exciting prospects will open up for the physics of high-energy particles when an accelerator with colliding proton and antiproton beams, which is now being built in Akademgorodok near Novosibirsk, is started up. In it will collide the particles that comprise the bulk of matter and antimatter. The energy of the particles in each beam will attain 25 billion electron volts. The reactions will be able to give birth to new particles and antiparticles — for example, quarks (if they exist), nuclei of antimatter, heavy mesons, and hyperons. Carrying out these reactions in the normal way would require an accelerator with an energy of 1,300 billion electron volts. The cost of such an accelerator would be more than a billion dollars, by international standards. The price for the facility being constructed in Novosibirsk is considerably less.

All this does not at all mean that the era of single-beam accelerators has ended. Many important experiments, including those with the mysterious "neutrino" particles, can be done only with such accelerators. However, there is no doubt that the future for physics at the highest energies is in colliding beams.

Therefore, it seems that it is time to change the present situation in which only a negligible fraction of the financial resources is being spent for making colliding beam devices.

The Institute of Nuclear Physics of the Siberian Division of the U.S.S.R. Academy of Sciences began building a device with colliding proton-antiproton beams under a very original financial arrangement. With some qualifications, it could be called the method of direct self-support for basic science.

In the course of our work on basic devices, we made intermediate ones: i.e., medium- and low-energy accelerators. We received permission to sell our devices as ready-made products, for a price not to exceed the price for something made for the same purpose. The money we made, we could spend to develop basic research.

An accelerator small enough to be transported in a car can produce, when it is switched on, radioactivity that is equivalent to that from many tons of radium, yet, when switched off, is absolutely safe and harmless. There has been a rapidly increasing demand for such devices. In three years we have received contracts that total 15 million rubles, and the scale of this work now exceeds the budget appropriations that the Institute receives from the Academy of Sciences.

The beam from an accelerator has turned out to be a good worker. For example, when irradiated, polyethylene becomes a superior plastic, which, while retaining its cheapness and technological properties and wonderful insulation features, also acquires resistance to high temperatures. Thousands of kilometers of cable have already been irradiated. Very soon some cable factories will be equipped with such accelerators.

At the Institute we have a three-centimeter thick sheet of steel on which a half-meter tall exclamation point was burned by an electron beam that had been brought out into the open air. It looks like an exclamation addressed to the metallurgists: "We offer you a new and powerful means for welding, cutting, and smelting!"

We are also concerned about the disinsectization of grain in storage elevators and granaries. All over the world there is an enormous amount of grain that is lost due to destruction caused by granary insects. However, it is possible to select absolutely harmless doses of radiation for the grain, which can halt the reproduction of these vermin. The first extensive test will be held this fall, in the southern part of the country. One does not need to be an economist to speculate on the overall value of the produce saved. It suffices to say that all capital investments will be repaid in one farming season.

Proton accelerators for an energy of 200 million electron volts also can have an important application. It has been proven both theoretically and experimentally that treating cancer with protons is much more effective than the X-ray and gamma-ray therapy currently in use. However, conventional proton accelerators for such energies are very complex, expensive, and unavailable for general use even at hospitals in the most developed countries. Therefore, the development of an inexpensive and easy-to-use proton accelerator is an important and noble thing to do.

A beam of accelerated particles is wonderful. With it one can search for minerals, sterilize medical supplies, can food, and disinfect sewage. With it one can transport energy for long distances, check the thickness of concrete and metal, and create new molecules and even atomic nuclei that are not present in the Mendeleev periodic table of the elements. Finally, only by means of this radiation is it possible to create antimatter and new elementary particles. It seems incredible, but it is true. But come to think of it, there is nothing strange about this. The extraordinary concentration of energy is what yields new qualities.

The brightly luminous beam from an accelerator, when directed into the open air, is surrounded by clouds of brown smoke. This is the air burning: it is a well-known process when nitrogen and oxygen combine.

The high efficiency and power of the new accelerators make the process of obtaining nitric acid from air and water economically beneficial. The day is not very far off when trains will begin to transport what the accelerator manufactures — thousands of tons of nitrogen fertilizers — from enormous cubes as tall as multi-story buildings.

But independently of however valuable these tons of "return on one's investment" will be, I think that what this 20-year old accelerator has produced that will fit inside a briefcase is nevertheless more valuable. This is so, not only because modern industrial accelerators are based on it, but also because this acquired knowledge contains so much for the future.

Buy an Accelerator!

G.I. Budker

As a rule, so far accelerators have been constructed for research purposes — for studying the structure of matter. However, their penetrating radiation has some great practical possibilities hidden in it. I will mention some of them.

The ability of particles to penetrate any obstacles, sometimes as much as several meters thick, is used in introscopy, or "interior-vision."

The ability of high-energy particles to excite and destroy molecules of matter, thus leading to the formation of new materials, has become the basis for a new promising area of science — radiation chemistry.

The deadly effect of specific doses of radiation on bacteria and insects can be utilized for the disinsectization and disinfection of grain, the sterilization of medical supplies, the canning of food, the disinfection of sewage, and so forth.

Originally published in the magazine *Ogonyok*, No. 19, page 17, 1969.

Radiation is of great assistance to physicians and biologists when they desire to stimulate processes that help maintain life and to suppress harmful processes in living cells. A well-focused beam of radiation that contains an enormous concentration of thermal energy can be used for cutting and smelting metal and for drilling through rocks. And, finally, it provides a solution to another very interesting problem — the transport of energy over large distances.

It is not at all strange for a new phenomenon discovered by scientists to have such wide-ranging practical applications. This is always the case. For example, it was like this with electricity. Could Faraday have actually imagined that very soon electricity would be used in homes, transportation, medicine, and metallurgy, for transferring speech and images, and, finally, in simulating processes that are similar to thinking?

Radioactivity was discovered decades ago, and radioactive isotopes have already been known for a rather long time. Why then has the broad use of this phenomenon in industry and agriculture begun to be discussed only now?

Budker: There are some reasons why things happen in this way. First, the use of radioactive isotopes is highly effective only when the radiation doses are small. But a device that was a powerful, safe, and inexpensive source for radiation just did not exist — and this is the second reason. Such devices have now become available.

There is, however, also a third reason, which I should call a psychological one. People have a fear of radiation, because the shadow of the hydrogen nuclear bomb hangs over this issue. Nevertheless, nobody of course will refuse the pleasure of watching television or shaving with an electric razor just because electric chairs are used for executions.

One must not confuse radiation and radioactivity. Radioactive elements that constantly emit radiation without being controlled are extremely dangerous. If a measurable, albeit small, quantity of these elements gets into the air or into the water supply, and thence into people's bodies, there are undesirable consequences. The accelerators that are built in our institute produce radiation only when and where needed. When not in operation, they are as safe as a switched-off X-ray device or transformer box. As for irradiated substances per se, including foodstuffs, they contain no induced radioactivity, and using them is as safe as holding in your hands an X-ray photograph of your lungs or stomach. The accelerators built in our institute are absolutely safe, inexpensive, highly efficient, and easy to use.

But how has it come about that it is your academic institute, which seems to be occupied with scientific questions very far removed from our everyday life, that began to work on problems that are essentially concerned with the national economy? What effect has this had on your main activities?

Budker: Industrial accelerators arose as a by-product of our work on developing devices for investigating the structure of matter and antimatter. Of course, it would have been possible not to divert people from working on our main subject. However, the natural desire of every inventor to apply the results of his work more actively won out.

It wasn't so easy, however, to persuade industry to take up this new line of work. And so we decided, by setting up two new laboratories at our institute without increasing the number of staff, not only to try to develop the design for these accelerators, which are so necessary for our country, but also to manufacture them and introduce them into the national economy ourselves.

Very soon we became involved in a very interesting economics experiment that mirrored the spirit of our time. The curious thing about it was that this activity not only did not contradict the main work of the institute, but it instead afforded the opportunity to address in quite a different way the questions of managing and financing science. The profit realized from building accelerators at our institute is spent on the development of fundamental science. Already we cover a part of the costs with these means, and therefore we can develop our investigations more rapidly and actively. This accelerator work required that large pilot production facilities be set up at the institute — such as machinery shops and design departments, which, in turn, made our scientific institute, which constantly needs extremely complex research instruments, no longer dependent upon industry and more flexible than before.

More and more laboratories are becoming involved in this new lifestyle. We build accelerators for the chemical industry, for agriculture, and for medicine — for example, for proton cancer therapy, which is more effective than gamma-ray therapy.

I want to dwell particularly upon one of the most important agricultural problems — the problem of disinsectizing grain. All over the world, a lot of grain, stored in granaries, is lost due to small granary insects. If, however, the grain, while being deposited in the granary, is irradiated with doses much smaller than the doses used for canned food, the reproduction of these vermin is completely halted. I think that the value of the grain that is saved in this way (just in one year!) will exceed all the expenses for academic science.

Do you think that all academic institutes should be involved in what you call this "new lifestyle," which implies "starting to become self-supporting"?

Budker: By no means! Scientists must not become slaves to their customers. Sometimes scientific research requires decades of unsuccessful work. Basic science in any country, and especially in our socialistic country, must be financed from the national budget, as is now being done.

But reasonable variations are also necessary. My assertion is that institutes that are now in a situation where they are capable of providing self-support for scientific research should be given the legal and practical opportunity to do so. This would be beneficial for the institutes, extremely important for industry, and useful for the state in general.

The issue that we are discussing is a nationwide one that is extremely important and affects the interests of a very large audience. Your magazine is one of the most popular and widely circulated publications in the country. I wish to utilize the two-million reader circulation of *Ogonyok* as a herald to declare once and again: the atmosphere of distrust related to using radiation in the national economy must be diffused, along with certain psychological obstacles. My wish is that directors of enterprises and institutions, engineers, physicians, geologists, and those who work in the food industry, will know that it is time for broad application of this new phenomena, and that it promises great benefits to our country's economy and to our people.

By Way of Antiplasma to Antimatter

G.I. Budker

One of the major tendencies in the development of modern physics is the drive to obtain higher and higher energies in charged particle accelerators, in order to increase the reaction energy of interacting particles. The scheme for such experiments has not changed since the time of Rutherford: a beam of fast particles bombards a fixed target. However, this scheme is very ineffective at high energies when the the speed of the particles approaches the speed of light. The mass of "projectile-particles" at such speeds increases significantly, becoming considerably heavier than the mass of the target particles. When a heavy projectile collides with a light particle in a target, only a small fraction of its energy, which had been attained with such great difficulty, is expended on the reaction itself. The lion's share is expended simply in the motion of both particles.

Originally published in the newspaper *Za Nauku v Sibiri* ("Pro Siberian Science"), January 14, 1970.

We decided to pursue another direction — to make the target move and have two beams of accelerated particles with equal energies collide with each other. In this case, the masses of both "projectile" and "target" are equal, and all of their energy can be transformed into interaction energy.

When the particle speed is close to the speed of light, the interaction of colliding particles is not four times more effective, as would follow from Newtonian mechanics, but many more times; this is a very important fact. For example, when two electrons collide that are flying along in opposite directions, each with an energy of a billion electron volts, the interaction turns out to be as effective as that in a standard accelerator operating at an energy of 4000 billion electron volts.

Actually, the idea of colliding beam accelerators is not new, and it does not involve any scientific discovery. It is simply the consequence of Einstein's theory of relativity. Many persons had suggested this idea before us, but as a rule they were pessimistic about the possibility of carrying it out. And this was quite natural. In fact, in a standard accelerator, the density of a "moving target" particle beam is a hundred million billion ("1" followed by seventeen "0's"!) times less than the density of a fixed target. The problem of having two particles collide is about as difficult as "arranging a rendezvous" between two arrows, one shot by Robin Hood from the Earth and the other by William Tell from the planet that revolves around the star Sirius. However, the advantages of colliding beams in comparison with traditional methods are so considerable that we nevertheless decided to overcome the difficulties. What was required was to increase the density of the beams and make them pass through each other many times.

Our "firstborn" had colliding electron-electron beams (which by now is ancient history for us) and consisted of two rings only 43 centimeters in radius; the interaction energy of the electrons was equivalent to that of a single-beam accelerator with an energy of 100 billion electron volts. No accelerator of the classical type can yet produce such an energy.

Our next one was a device with colliding electron-positron beams, each 700 million electron volts. Nowadays it is used for experiments to create new particles from the electron-positron annihilation process — viz., the electrons of antimatter. The process of systematic studies of matter and antimatter apparently began from this device. A similar accelerator was started up a year later in France, and this year another in Frascati, Italy. In 1970, facilities like this, but with considerably higher beam energy (almost 3.5 billion electron volts) will begin operation at Stanford University (in the U.S.A.) and at Novosibirsk. Thus, colliding beam accelerators have not only earned their right to exist, but have also become fairly widely used. However, their real competition with classical accelerators began from the time when devices using heavy particle beams — i.e., proton-proton beams or, even more interesting, proton-antiproton beams — were constructed. The point is that for the case of light particle collisions, even the smallest device, as has already been mentioned, exceeded the limits of potential of traditional accelerators from the point of view of interaction energy, and hence this line of research has opened up a new area in the high-energy physics of light particles.

Meanwhile, most problems in modern high-energy physics are related to heavy particles. Here, the energy range of the colliding beams still overlaps with the classical domain.

The European Center for Nuclear Research (CERN) in Geneva is constructing two large storage rings for colliding proton beams. The famous CERN synchrotron will deliver particles to these rings. The results from the colliding beams will make this device equivalent to an accelerator with an energy 50 times more than the energy of the synchrotron. The construction of a device that produces the same energy, but contains protons and antiprotons — the particles that constitute the bulk of matter and antimatter — is now underway in Novosibirsk.

Naturally, we are ardent advocates of the new method. However, we should not forget that classical accelerators, when bombarding dense targets, produce intense beams of secondary particles that are necessary in a whole number of important physics experiments.

Colliding beam accelerators and classical ones are not mutually exclusive, but complement each other. If the question of the cost of accelerators played no role, perhaps only classical accelerators would be built.

In the energy range of several thousand GeV (billion electron volts), traditional proton accelerators can still compete with colliding beam accelerators; however, super-high energies are only the domain of colliding beams. Therefore, Siberian physicists are now discussing a project for a device with colliding beams of protons and antiprotons that would be equivalent to an accelerator with an energy of two million billion electron volts. Such an accelerator, if constructed in the traditional scheme, would have a diameter exceeding that of the Earth, and its price would be greater than the gross national product of the whole world.

The Soviet Union has an undisputed lead in the competition with other countries in the area of colliding beams. We have in Novosibirsk approximately as many operating accelerators with colliding beams and ones under construction as the total number in all other countries combined.

From among all of the devices that we are now building, the one of most interest is the accelerator with colliding proton-antiproton beams, each with an energy of 25 billion electron volts. This is equivalent to a standard accelerator with an energy of 1,200 billion electron volts. With its help, it will be possible to obtain all the known particles and to search for new ones, whose mass may be approximately as much as four times the mass of the particles currently being created at the largest accelerators. If quarks really exist — quarks are truly elementary particles from which, some theorists suppose, all the other "building blocks" of the universe are formed — and if the mass of quarks does not exceed 25 times the mass of the proton, they could be discovered with this accelerator.

In addition, this new accelerator will be an original hybrid, in which it will be possible not only to accelerate heavy protons and antiprotons, but also to have electron and positron beams accelerated to six billion electron volts collide.

The experiments in this device with light particles will be extremely important for checking quantum electrodynamics. It will be possible to define the limits of applicability of this the only logical theory to date that explains most physical phenomena.

We are also going to carry out what is perhaps not the most important, but still an extremely interesting, experiment with antimatter. As is known, quite recently its existence was generally thought to be doubtful. Then, anti-electrons (i.e., positrons) were discovered in cosmic radiation. Thereafter, with the help of large accelerators, anti-hydrogen nuclei (i.e., antiprotons) were obtained. In Novosibirsk, on the VEPP-2 device, we obtained the first visible antimatter: it shone brightly and could be seen with the naked eye. The positron beam lasted for hours. This was already something real and perceptible, not just for physicists, but for anyone. Please take a look, here it is — light from antiparticles!

The device with colliding proton-antiproton beams will allow us to store about 10 billion antiprotons in the ring. This is a large number.

We want to attempt to create antimatter in the laboratory — this means not just producing antiparticles, but anti-atoms. To carry this out, we plan to make the positron beam propagate alongside the narrow antiproton beam at the same speed: then there will arise an antiplasma of antiprotons and positrons, in which anti-hydrogen atoms will be born. We expect to produce a high enough anti-hydrogen current that it will be able, for example, to burn through a sheet of paper. Thus it will be possible to study the properties of anti-hydrogen and, in particular, to investigate its spectrum. According to all the theories, it should be no different from the spectrum of usual hydrogen, but no experimentalist would miss the opportunity to check this assertion. Should they be different, it would cause a real revolution in our ideas about the nature of matter. At present, astrophysicists are debating whether anti-galaxies exist in the universe and whether matter and antimatter stand on an equal footing. Perhaps our experiments will be an "arbitrator" in this debate.

Good Student–Good Scientist

G.I. Budker

The answer of academician A.M. Budker in response to a letter to the newspaper Izvestiya *from a student, Yu. Smirnov.*

Yu. Smirnov's letter:

I want to ask a question: Is it necessary for students to be excellent? What is meant by an "excellent student"? This is someone who has perfectly assimilated the curriculum. He just has no time to do anything else. Hence the expression: An excellent student must, to some extent, be lazy (since he does nothing that is unnecessary, otherwise he would have a breakdown). It is difficult to be an excellent student.

Originally published in the newspaper *Izvestiya*, February 10, 1972.

But does it make sense? Such a person has practically no advantages, even when entering the university. There, excellent students are regarded as careerists, and rarely does one of them get a mark of "5"[2] on the entrance exam — they get "cut down."

Then, the excellent students go to universities. Basically, these are the people who are going to become scientists. I don't know the statistics, but it seems to me that such people do not constitute the foundation of modern science. In my opinion, a mark of "3" along with "5's" is a better proof of talent than straight "5's."

What about a diploma *summa cum laude*? This, however, doesn't mean any real advantages or enhanced opportunities.

I myself was an excellent student for fifteen years; I finished school with a gold medal; and now I am graduating from an institute, where for all five years I have never received any mark less than "5." Too late, however, realization has dawned upon me. Too late I have understood that nobody needs it. I should not strive for "5's," but instead develop my abilities in one specific area. "To know something about everything, and everything about something" — this should be the principle to follow in developing one's abilities.

Excellent students are few, just because nobody needs them. Science needs talented people who are not confined within the narrow limits of studying in order to excel. Therefore, one should not encourage getting "5's." It is knowledge that should be encouraged.

Yu. Smirnov, student, Kazan'

Academician A. Budker's response:

The question about evaluating knowledge is only a small part of the larger problem of today: namely, whom to teach, what to teach, and how to teach it.

The writer of this letter was undoubtedly motivated by good intentions to help improve education. However, I also discerned in this letter what you might call superfluous practicalism. Evidently, such an approach reflects the situation in real life, although in this case we must not forget that, when speaking about the powerful stimulus of financial

[2]Trans. note: In the usual Russian grading system, a mark of "5" equals an "A," "4" equals an "B," and so forth.

considerations, we should never encourage preoccupation with material well-being.

The writer of this letter asserts that bad students become good scientists. I don't know about this. There are no such people among my acquaintances. In the best cases, bad students can become fair scientific administrators.

It seems to me that there is a connection between how successful a student is now and his future as a scientist. Most good scientists were good students at their schools and universities, and the percentage of good students who become good scientists is rather high. Many more good scientists came out of a thousand excellent students than out of a thousand average ones. Such a connection, even though it isn't always a law of nature, certainly exists. Mr. Smirnov is now involved in a polemic with academicians M. A. Lavrentyev and S. L. Sobolev[3] — the most famous scientists in their area of knowledge. Sergei L'vovich Sobolev always had an easy time with studies and got all "5's." Mikhail Alexeevich Lavrentyev didn't get into secondary school because his ability in foreign languages wasn't always so good. Nevertheless both of them were perfect in physics and mathematics. And what about the experience of my contemporaries? Many of them, who are now talented scientists, could not have become physicists if they had entered Moscow State University a few years later than they did, when excessively strict requirements for composition were included in the entrance examination (for all of the departments, without exception). Moreover, Alexander Sergeevich Pushkin, who had unsurpassed talent and intellect, was poor in mathematics....

Now, something about the connection between good grades and good knowledge. There are some young people who have perfect knowledge, but who don't strive for good grades: they don't always do their homework and other assignments, and they don't always give answers successfully. There are also some young people whose knowledge is very average, but who strive for excellent grades by any means. (Incidentally, school children and university students usually have very keen discernment about the quality of a grade of "5.") And nevertheless there is a strong and clearly detectable connection between excellent knowledge and excellent grades. The stronger the living association between the

[3]Editors' note: Academicians M. A. Lavrentyev and S. L. Sobolev also published articles in the same issue of *Izvestiya*.

teacher, the educator, and his students, the stronger is this connection. With an intelligent and experienced mentor there is attained a sufficiently high degree of correspondence between the grades that are given and the real order of things. In each specific case such a teacher can discern whether a student is aimed at gaining knowledge and abilities or just making a career of getting good grades.

One thing can certainly be said: If the education is formal in character, the grades will also be formal. On the other hand, however, teaching that is creative will, as a rule, guarantee that grades will be correct and individualized.

I was a student at school during a time of great experimentation. We had a two-grade system ("satisfactory" and "unsatisfactory"), a three-grade system (with "very satisfactory" added), then a four-grade, five-grade, and even a hundred-grade system. Even now I still remember one of my grades — 97.3% — but I have never understood what considerations were the basis for the 0.3% part. In spite of all the variety of grading systems, we did not see any practical difference in the results for acquiring knowledge. To tell the truth, though, the introduction of the team system (in which the grades of the whole team depended upon the answers of one of its members) afforded me the unexpected opportunity to get excellent grades in German, but a "2" in mathematics.

Even though it is very important for grades to be correct from the point of view of training, their role is incomparably more important when they determine people's futures. Then they acquire social significance. When the students of Novosibirsk State University are assigned to jobs, it is possible, just as at some other graduate schools, to overcome the consequences of possible noncorrespondence between grades and abilities. During their final three academic years, the young people do their training work at academic institutes, and the jobs they obtain depend on the results of this training work. In this case the grades they receive in their examinations play an auxiliary role.

For the same reason, it seems to me that the entrance examinations for universities should not be competitive. They should be replaced by student competition during the first three academic years.

As for the thesis "something about everything, and everything about one thing," this, it seems to me, is becoming more and more powerful today. The emergence of persons with encyclopedic erudition, like Leonardo da Vinci or Lomonosov, is now impossible. Sciences initially intersect each other where there are common interests and collaboration among scientists. A new area of science can arise at such an intersection, and then experts with narrow specialities again become necessary for its development. For example, a biochemist is not someone who has perfect knowledge of both biology and chemistry, but rather someone who specializes in biochemistry.

It seems to me that Mr. Smirnov wishes to receive precise directions about what he should do. To search for this is the same as searching for the philosophers' stone. If a specific formula needs, however, to be stated, it would go like this: Be reasonable, be tactful, be benevolent, and have a deep sense of responsibility.

Expected Unexpectedness

G.I. Budker

And now some questions for the director of the Institute of Nuclear Physics of the Siberian Division of the U.S.S.R. Academy of Sciences, Lenin Prize laureate and Academician, Andrei Mikhailovich Budker.

What do you foresee for the future of technical progress and, consequently, of civilization on our planet?

Budker: The future is naturally contained within our present civilization.

Originally published in *Nedelya*, a weekly supplement to the newspaper *Izvestiya*, Issue No. 1, 1974.

Technical progress, which began about 500 years ago, nowadays proceeds at a particularly fast rate. Therefore, I think that all sorts of difficulties arise, such as an energy crisis, a food crisis, environmental pollution, and so forth. But isn't this similar to the fears that a pithecanthropoid man would have if he were shown our present-day lifestyle? Imagine: squeezing your feet into tight shoes, refraining from climbing trees, locking yourself into something like cages called apartments....

I am well acquainted with the opinions of foreign experts, such as those published in this same issue. I can't agree with them for the following reason. We must not confuse progress *per se* with the costs associated with management that is self-interested and unrestrained. The energy crisis, currently spreading in the capitalistic part of the world, must be considered precisely such a cost. There is no cause to doubt that with a reasonable and planned economy, it would be possible to avoid both an energy crisis and environmental pollution for an unlimited time, even without the employment of atomic energy, with only the use of old, known energy sources — water, wind, solar, coal, and oil. But all this could only be possible if technical progress were delayed. Moreover, social progress would also be delayed.

Atomic energy, apparently, is the only source that guarantees absolute opportunities for energy and, together with this, absolute opportunities for the subsequent unlimited development of industrial progress.

Discussions about polluting the atmosphere and damaging the Earth by the use of nuclear and atomic energy only concern the case when these energy sources are used without control or thoughtlessly. When the solution to this problem is thought out and carefully elaborated, there are no significant difficulties with utilizing the waste products from these sources.

We are unable to do anything with the trillions of cubic meters of carbon dioxide that are produced by burning oil and coal and that will slowly transform the Earth's atmosphere to one like that of Venus. However, I'm sure that utilizing the waste products of atomic energy, even if subsequently it is very widely applied, will not be an insoluble problem for us. The energy content of atomic fuel is a million times greater than that of ordinary fuels, and its waste is a million times less.

The Earth has enormous resources of atomic energy, sufficient for many hundreds of years. And as for thermonuclear fuel (which is nothing but water), even in the case of unlimited development of progress on Earth, it will be practically sufficient to last for millions of years. Thus, the problem of energy sources will no longer be one of the problems of mankind.

Thermonuclear fusion energy should begin to work in the next few decades. The fact that it isn't already being used is somewhat of an absurdity. This is a rather rare case in physics: scientists have worked hard on this one problem now for 30 years and have not yet solved it. But they will solve it — and they will do so in the near future. Moreover, there is the hope that at the same time the fusion problem is solved, physicists will also solve the problem of directly converting fusion energy to electrical energy, since a high-temperature plasma, which is needed for thermonuclear fusion to occur, is an ideal object for it.

Mankind will acquire a practically unlimited source of electrical energy. People will become all-powerful.

Andrei Mikhailovich, in your dreams how do you imagine the Earth will look, say, in the middle of the third millennium?

Budker: I connect the future of the Earth with the development of atomic industry. The reason is because the atom is not only an unlimited source for energy, but also a powerful force for transformation. All atomic industry involves dealing with enormous amounts of energy. It will be especially needed in modern "alchemy" — i.e., in obtaining new elements. Each substance that is obtained by means of this nuclear "alchemy" will require an energy expenditure that is a million times more than today's chemical processes. However, all the heat that arises in this case will remain on Earth and can lead to global overheating, to the most severe consequences that are even difficult to predict.

I'm sure that at that time, in order to prevent these things from happening, society will move the atomic industry to outer space or to other planets. (Space flights, including cargo flights, to other planets in our solar system will become quite commonplace in those days — atomic energy automatically solves this problem, too.) One can imagine the following idealized scenario: all production is located on the Moon, or on Mars, or directly in space. Even agriculture, if necessary, could also be carried out in space — then it would not depend on the weather and would be completely controlled by people. The Earth would be the common dwellingplace for mankind. I can hardly believe that man, with his physiology, would feel as completely comfortable elsewhere as he does on Earth, with all of its storms and winds, yet also with its wonderful sunshine and wonderful plants. The Earth will reacquire its original appearance: fields, factories, and industrial plants will disappear from its surface, and it will become the place where people will live, relax, and enjoy themselves; industry, including the food industry, will be located outside its boundaries. Then the Earth will be able to support an enormous number of people. As a dwelling place it is immeasurably large.

The Importance of Scientific Schools

G.I. Budker

The question "Should a scientist have pupils?" is rather artificial. It is the same as asking whether people need to have children. They are the ones who continue the work we began and bring it to a logical completion. And if they themselves do not finish this work, it will be completed by their pupils. Actually, this is the way of scientific progress. The teacher achieves immortality through his pupils, just as every man becomes immortal through his children.

Originally published in the book *The Age of Knowledge* (Molodaya Gwardiya, Moscow, 1974), pp. 124-142.

Even for a very talented man, it is quite difficult to do anything in modern science without assistants — and pupils are first of all assistants. But this is not the whole point. When rearing children, as a rule we do not think either about the continuation of the human race or about the preparation of support for our old age. Likewise with a scientist: when fostering a pupil, he behaves following his instinct, which is similar to the instinct of reproduction. He is truly happy even when his pupils leave him for an independent scientific career — as long as they are good scientists.

It is unnecessary to prove to someone who is entering science how important it is to have a kind and intelligent mentor. If asked, every scientist will always remember the person to whom he is indebted for having encouraged his initial fledgling interest in learning and for having given him kind advice when he was choosing his first scientific project, without which it is impossible to learn to overcome difficulties, as well as for doing a great many other things, without which no one is able to become a researcher.

To study only from textbooks, monographs, and papers is the same as trying to master the secrets of the art of playing the piano from a self-instruction manual. Even for me, someone in the exact sciences, it is difficult to explain why this is impossible, but nevertheless it is true. Writing notes of music and jotting down where the music should be played "forte" and "piano" is easy, and representing with symbols lots of other details and pecularities of performance is possible, but attaining to a high professional level without a teacher is impossible. The same is true in science: It is impossible to master the secrets of the skills for doing research without a good school. Even though in developed countries all the current scientific literature is available to practically everybody, it is not accidental that good physicists arise where there is a good school. Of course there are some exceptions, where a person is indebted only to himself for his principal success; but these, according to the common expression, only confirm the rule. Occasionally, concerning a talented man, it can be said, "Yes, he is talented; however, he has not had a good education."

Understanding what constitutes a scientific school — in particular, a scientific school in experimental physics — involves not only understanding some complicated world-perspective ideas of modern physics, not only knowing the group of problems and objectives that are most expedient to work on, but also understanding a whole host of small elements of work, things that are used daily, even hourly, but that are not — and simply cannot be — described either in textbooks or monographs. Suppose, for example, that when the huge chamber of an accelerator device is being cleaned, a small thread of gauze, invisible to the naked eye, were left there. During an experiment, particles will be lost here. You will obtain strange effects when adjusting the accelerator. If you are intelligent enough and have learned from textbooks all the theories about particle motion in an accelerator, you will jump to the conclusion that a so-called resonance has occurred. Repeated experiments will seemingly confirm this and will even give the number of the resonance. Over and over you will open up the chamber and change the magnetic field, struggling against this superfluous resonance. Yet it will continue, until at last you accidentally brush off this thread, without noticing it. And then the accelerator will work.

However, if you have a teacher, he, being a skilled experimentalist, would certainly tell you sometime about such cases and how to recognize them. There are numerous much more profound things — I gave the simplest example since experiments involve not only science, but also a great deal of art, and so far no one has been able to learn art from textbooks.

Training young researchers is also similar to an art since it requires of the teacher skills and methods that are truly artistic.

What should a young scientist begin with? Which problems should he solve?

If a junior mountain climber is sent straight off to conquer high summits, he will certainly break his neck. However, if his trainer always gives him overly simplified assignments, his skills would diminish so much that he will never be able to conquer a more or less significant height. Here is where a teacher displays his art for sensing and understanding the measure of training, the precise magnitude of a task — by means of their difficulty and nature.

I have met many people in science who had overstrained themselves due to tasks beyond their ability, which had been given to them by negligent teachers and mentors. Many people lost their way because they lacked scientific perspective and wasted themselves in small, unimportant problems. And I am sure that most of them were really talented people who were just unlucky with the teacher they had.

Of course, it is the teacher's privilege to chose the precise intellectual burden. Weightlifters know how important the timing is when requesting the appropriate weight and also the exact calculation of the muscle loading. Their trainers naturally help them with it. But also the pupil, like a weightlifter on a stage, must remember that it is always necessary to tailor the tasks one undertakes to one's abilities.

Naturally he very much desires to work on the most difficult problems that no one else has been able to solve. Then a certain paradox occurs, as follows, which leads to self-deception. Both the young scientist who has just recently entered a new area of science and the famous scientist who still has not been able to solve given problems are, in some sense, on the same level. As long as both of them have no results, both of them are formally equal. (Two zeros are always equal to each other.) Thus the young inexperienced scientist, whenever he gets practical opportunities to speak at seminars and scientific meetings on a par with everyone else, feels that he is included within the circle of great people. They object, argue, or agree with him. These discussions "on the same level" continue only until the moment when certain important details that can open the way for solving the problem are made clear. This is the moment when experience and skill exert an influence.

But scientific modesty, when one thinks that he must work only on small applied problems, is also detrimental. It deprives the scientist of his outlook and, consequently, of the opportunity to work on big problems in the future.

How then to combine these two principles, seemingly mutually contradictory, in the training of young scientists? It seems to me that in this case, both teacher and pupil must remember that science is not only art and poetry — it is also a trade. The teacher must teach the pupil how to think poetically, because the poetry of science is its acme. However, as with all other educators, he must also teach professional skills. From time to time it is necessary for the pupil to obtain concrete results, so that his abilities and qualifications can be judged from these results.

In general, it is difficult to give a specific prescription for behavior, what a physicist should be doing and how. One's actions should correspond to one's task. Besides, each scientific problem becomes the subject of personal interest for the one who works on it, and in terms of hopes for success, a problem that requires ten years of work costs ten times more than a problem that requires one year.

Now let me say something about fashions in science. Fashions in science can be either justified or not, unlike fashions regarding the length of skirts or the height of shoe heels, for which there are no objective criteria. Nuclear physics, genetics, and space research are the greatest fields of science in our modern age. They have wonderful futures, upon which the fate of mankind depends; they have content, methods, traditions, and schools. Such fashions are justified.

However, in history, as well as in our own day and age, we have known sciences, quasi-sciences, and pseudo-sciences that became extremely fashionable at a certain stage and then died out, without acquiring any content. For example, let us recall astrobotany (essentially the science of the plant life of Mars and Venus). It is really possible that some day plants will be discovered on other planets. But 20 years ago, when astrobotany arose, it was impossible for this kind of science to exist. In the best situation it could refer to separate studies in astronomy or botany.

Prematurely declaring some individual results from established areas of science to be an independent new area of science is as dangerous, in essence, as when a youngster begins to live independently outside the family circle. Of course there are some advantages, but they are much outweighed by the shortcomings and dangers. The main danger for areas of science that "hatch" prematurely is the absence of criteria. An area of science in which the criteria for correctness and significance are poorly developed will be vulnerable to and unprotected against ignoramuses and rogues.

Thus, nowadays when such fashionable "sciences" cause a sensation and when a certain amount of publicity about them is generated — in such cases considerable number of talented youth who have not yet found their own way are disoriented. Most of them will never become scientists because instead they go into those areas where there is neither anything for them to do nor any problems that are simultaneously important and valuable and, in principle, solvable within our lifetime. Therefore the agitating of youth must be regarded very seriously, and one must be very careful not to make young people embark on a search for beautiful words, dreams, or hopes that can hardly be realized. Even though science involves dreams and hopes and poetry, nevertheless they are specific dreams and specific hopes. It is important to understand what science consists of, what are the necessary elements of science, what subjects must be included in an area of knowledge, and what is required for this area to be designated as "science."

Here is an example from a field that I am close to. Many scientists all over the world are now working on the creation of controlled thermonuclear reactions. At some time (obviously in the near future), controlled "fusion" will be achieved. However, this area of science — "controlled fusion" — does not exist, even though numerous scientists all over the world are occupied with this problem and much money is being spent on it. As time goes on, it will become a branch of science that will hardly have to defer to other fields of science or of the national economy such as electronics, thermal engineering, and many others. But at present it is an area in physics. If someone thinks it is possible to first create a profession and to teach people (teach them what?) and then to have these people form a science, he is, in my opinion, mistaken. Thus, it is not enough to prepare a name for an area of science (or even to fill this name with a certain significance) in order for this area of science to begin to exist. Deeper grounds and prerequisites are needed for this.

Perhaps the comprehension of this fact will keep many famous scientists from making speeches in public in which they promote directions that have no promise (for young people). And, on the other side, pupils themselves — present as well as future ones — must cast a critical eye on this sort of propaganda and must try to penetrate deeply into the subject that constitutes the field of their future scientific activity.

For a young man who is thinking over his future and making plans, it is important to know what the present areas of science are in general, which ones are flourishing, which ones are developing rapidly and which ones more slowly, and which areas of science are expected to produce the most significant results in the near future. It is important for a young man to know which scientific directions will become actualized — not only from the standpoint of society's needs, but also from the standpoint of opportunities within science itself.

How young scientists are distributed within the sciences must be adequate to the research fields in each of these sciences. Each task and each development has a required scope of activity. Gathering one million people to build a single house is impractical — they simply would have nothing to do, not to mention that this is obviously unreasonable. Those who have selected science as their profession can encounter one more danger, which is probably even greater than the one already mentioned. In recent times, it has become more and more difficult for the uninitiated to distinguish "science" and "science-like things" — more precisely, who is a scientist, and who is a "scientist-like person." Not every Candidate of Science, not every Doctor of Science, not every person with a scientific degree is a scientist. To the contrary, occasionally there are real scientists who, due to circumstances, have not obtained lofty scientific degrees.

Can a criterion be given that would make it easy to tell a real scientist from a hypocrite and to distinguish real science from things that pretend to be science? I think it can hardly be done in one sentence, let alone many sentences. Scientist-like people feign to be scientists with such adroitness that only an experienced eye can spot them as frauds. They adopt scientific terminology and phraseology. Of course one cannot deny that they have a good feel for novelty, especially if these new things come from an acknowledged authority. It is as difficult to spot them as it is for amateurs to tell the difference between good music and bad music.

A natural criterion for telling the difference is whether the work has an effect. Science gets results, but "science-like activity" has none. But again, only top-flight experts can perceive this effectiveness, because "people near science" can also create the appearance of results. This hides another enormous problem for young persons.

Apparently it is impossible to state with absolute certainty how youth can be taught to tell the difference between science and "science-like activity." Modern science is so complicated, that in fact the youth are helpless in the face of such a choice. In this case, as I have already stated, some severe requirements must be placed on those persons who have a reputation for being authorities and are known as famous scientists. The orienting of young people and their scientific interests depends, in particular, on the public speeches of such persons. But we must not forget that pseudoscientists also make use of the mass media and sometimes cause a pseudoscientific sensation concerning unimportant and occasionally even vicious ideas. The greatest evil caused by these pseudoscientists with titles and degrees is not even the fact that they waste financial resources; rather, they are wasting one of our main riches — national intellectual resources. Having captivated talented youth, they corrupt them ethically (for example, by implying that success in and of itself is important, independent of what it is based on), or, if they fail to corrupt them, they deprive them of the scientific world-perspective, of understanding what is true and what is false in science.

It seems that our youth, first of all, must receive good moral and ethical training. This should help even young people to distinguish scientists from pseudoscientists.

Youth must be able to distinguish words from actions, truth from falsehood, primitivity and lies from true depth and imagination. They must know that science is not a sterile world where people's ethical behavior is ideal. The world of science is the same as that of real life, where there are plenty of both rascals and bystanders and also heroes, where noble and wonderful characters exist side by side with those who are banal and careeristic and just plain swindlers. However, if the youth are warned about it, they will learn to discern all of these vices of the scientist-like "authorities," whatever hypocritical robe they might be dressed up in.

Fiction, which always teaches people the discernment of good and evil, can play an extremely important role in this sort of training. Unfortunately, neither in our country nor abroad are there yet any writings in which these kinds of problems have been considered in any depth.

It is also important for young persons to understand that if they encounter banality, dullness, or narrow-mindedness, it only means that he (or she) has just been unfortunate. They must know that indeed there are really bright, decent, and very intelligent people in science. A young man may be disappointed just with his scientific leader, but he must not be disappointed about science itself. He should understand that he has made a mistake, and the faster he corrects it, the more chance he will have of finding his own "temple of science."

There is a well-known expression: "A man is a fraction, the numerator of which represents what he is, while the denominator represents what he thinks about himself." I would say that a scientist is a fraction, with the numerator being who he really is, and the denominator being an average among what he thinks about himself, how he behaves and represents himself, and other things, which can be generally referred to as "decency." If one thinks more highly of oneself than he is in real life, he can hardly be a decent person. If he knows that he is nothing, but pretends to be someone important, he is quite indecent. But, if on top of all this, he does something disgraceful, then he is absolutely unethical. Therefore I would define the formula for human dignity as being a fraction whose numerator is an objective evaluation of the person and whose denominator is related to his decency. This word seems to be somewhat lost from our language. The words "honesty" and "principles" are there, whereas decency is a concept that cannot be precisely defined, but which involves all the positive ethical characteristics of a person: honesty, strong principles, reasonableness, tolerance, and many, many other things. Moreover, all of these characteristics must come naturally for a decent person.

By and large, famous scientists are decent persons. A decent person will never occupy the place in science that does not belong to him — someone who occupies the place that is not his own is unethical. Here I wish to emphasize once more the importance of schools of science as the environment for the true understanding of scientific problems and scientific ethics. I am sure that a scientific group in which ethical principles are violated is lost for science, even though it can exist as an institution. Genius-villains in big science occur only in bad science fiction novels.

Now a few words about choosing a scientific profession. Each young man must choose from among the fashionable and unfashionable sciences one that corresponds most closely to his abilities and calling. He must, however, choose a real science — not a pseudoscience, nor a nonexistent one. When selecting a field of activity and its topics, everyone must make for himself a very definite decision, since human life is short, and all digressions from a subject are very costly — they irreversibly consume many years. Understanding that science, even at the highest intellectual level, is work and not entertainment will help to avoid this. And, even when it seems to be entertainment, it is entertainment in one's work.

In spite of everything I have mentioned, I continue to remain optimistic. I know, because reality convinces me, that nowadays youth are also basically choosing both the fields for their scientific endeavors and their teachers in science correctly.

These days it is often said that young people have lost interest in physics problems, in particular in the problems of nuclear physics. I think this is wrong. To the contrary, nowadays nuclear physics is being liberated from many people who had previously joined it as something popular. However, it is a fact that a part of the talented youth are moving into the field of molecular biology and genetics, where great events are expected to happen. As for myself, I very much approve of young people being interested in medicine, which is again becoming a "man's profession." But nevertheless, so far the best and most talented youth go into theoretical and experimental physics. I can see it for myself at our own Novosibirsk University, and in general my entire experience of associating with young scientists has shown that, just as before, the most talented ones choose nuclear physics. However, it is possible to understand in what connection there appeared these statements about the decline of nuclear physics, statements to the effect that all of the great discoveries to be expected from the present generation of scientists have already been made.

All great discoveries can be classified, in correspondence to the "spirit of the time," into three types: namely, timely discoveries, belated discoveries, and premature discoveries. As examples, let us take the three great discoveries of our age — space exploration, lasers, and nuclear energy.

The exploration of space can be said to be an example of a timely discovery. Psychologically, mankind was quite ready for it, with the considerable help of science fiction that already (from the last century) had developed this subject down to the smallest of details. Rather serious predictions are confirming the opinions of the science fiction writers. Technologically, civilization was also ready for space flights: airplanes had been developed for a long time; then there came jet air planes; then missiles; then ballistic and intercontinental missiles; and finally satellites. Mankind was ready for it, both technologically and psychologically, and step by step it went into space. This is an example of a great discovery that was timely.

Lasers are an example of a belated discovery. The theory for lasers, i.e., the theory of stimulated radiation, was developed in the beginning of this century. Already in 1941 I was given a question about the theory of stimulated radiation on a degree examination at Moscow University. The psychological and practical need for lasers ripened long ago. It was expressed in *The Hyperboloid of Engineer Garin*[1] and in the death

---

[1] Trans. note: A science fiction novel by the Russian-Soviet author Alexei Tolstoy.

rays in the hands of the characters created by H.G. Wells in *War of the Worlds* and in other fictional examples, as well as nonfiction. In the 1930's, the physics of optics was developed practically to its modern level. The experimental techniques and the series of experiments that had been completed before that time were sufficient for constructing the first lasers. There was no need for a special industry to do it. Thus, everything — needs, industry, and science — were ready for the appearance of lasers on the eve of World War II. But it did not occur. Probably the war and then the work on atomic energy distracted the attention of the most famous scientists and industry from the problem, which obviously was unfortunate.

Without a doubt, an example of a great premature discovery is the discovery of atomic energy. Not long before the fissioning process of uranium nuclei was discovered — i.e., before the discovery of the possibility of using atomic energy — academician Abram Fyodorovich Ioffe, an extraordinarily progressive scientist, who perhaps can be called more of a dreamer than a sceptic, had asserted that the question of the practical use of atomic energy could only be raised a hundred years later.

Society was absolutely unprepared to use the opportunities of atomic energy. Even in the science fiction literature of the pre-atomic era there was no hint about the use of nuclear energy or, in general, about the internal energy of matter. There were only hints about the use of radioactivity. Science was absolutely unprepared for it. There was no theory for the atomic nucleus. Incidentally, there is still no theory for nuclear forces. Thus, in the usual sense, there was no science concerning atomic energy similar to those for lasers and space exploration. Technologically, industry was also absolutely unprepared to solve the problem; everything had to be started from the very beginning. Moreover, society did not yet need atomic energy. Nevertheless, atomic energy was born. That this process was, in principle, unnatural was also proven by the great expenses associated with the solution of scientific problems for obtaining atomic energy. For the first time in the history of science, these expenses became comparable to the annual gross national product of the most highly developed countries in the world! Clearly, the economical solution of these problems was also premature. One can say that World War II caused "premature childbirth": atomic energy appeared several decades earlier than it was supposed to. However, the child was born, it survived, and it began to grow up, not on a scale of days, but of hours. Already for some time now, atomic research has proved its worth from the scientific, political, and merely economical points of view.

The most important fact, however, is that this discovery of atomic energy caused a revolution in all other fields of knowledge. Never before had anyone dealt with problems of such magnitude. Even if someone had encountered them, he had retreated from the difficulty of this complex of problems, for which it was not clear how the priority and the significance of the problems should be classified, what should be considered first, what must be financed, how the results could be summarized, and whose opinion should be decisive about projects costing billions. The fact that atomic science exerted a revolutionizing influence on all scientific directions — organizationally, psychologically, and technically — was the precise reason for the fairly easy subsequent success in mastering laser techniques, space research, and many other things.

Now, when from time to time someone says that interest in atomic problems has considerably diminished and that the development of modern nuclear physics is decreasing, I think that those who say this became physicists or, in general, scientists only by accident. People who work in science must work extensively and diligently, as if they are preparing the fruits of a new harvest — not as if they were merely picking mushrooms after it has rained. Also, those who say this are extremely short-sighted.

To date, atomic science has continued to develop at an extraordinary pace. To prove this, it suffices to look at a graph of how the energies available to modern "nuclearists" have increased, since this is the reason that underlies the success of all experimental works in nuclear physics. Cyclotrons in the 1940's produced particle energies of 10 million electron volts. The synchrophasotron in Dubna, built in 1949, produced 500 million electron volts, and the so-called cosmotrons yielded billions of electron volts. The large accelerators at CERN (in Switzerland) and Brookhaven (in the U.S.A.) produce 30 billion electron volts. Serpukhov gives 70 billion; Batavia (in the U.S.A.) 500 billion. The energies in the reactions of elementary particles in colliding beams are even higher. And all this has been obtained in three decades. To give an idea of the scales involved, I need only remind the reader that the kinetic energy of a molecule (a particle) at a temperature of

10,000°C is only one electron volt. And the higher the accelerator energy, the more deeply we can understand the essence of matter and learn about the laws of its structure. After all this, can it be said that atomic science has come to a standstill? Of course not.

To a certain extent it has halted, when compared to the great outpouring, as if from a cornucopia, of wonderful practical results in its initial years. The atomic bomb, atomic electrical power stations, tagged atoms, atomic powered icebreaker ships and submarines, the use of artificial radiation in the national economy — naturally, these all aroused admiration. Very soon, in the next decade, we shall witness the mastering of thermonuclear fusion energy. Thereafter the application of the energy of antimatter (a fuel that is a thousand times more effective than nuclear fuel) is expected. If for several years there has been no news about some sensational practical results, by no means does it imply that scientists have obtained nothing.

Mankind could actually become exhausted by these incredible efforts and "need to take a rest" before again continuing to advance to the ultimate goal. But even this has not occurred. As a science, nuclear physics has not stopped developing even for one day. There is nothing comparable to the discoveries of recent years. A whole series of new particles, the absence of so-called C-symmetry and CP-symmetry — these are rather fantastic, incredible results. The present day and age can be compared to the eve of the discovery of quantum mechanics, when enormous quantities of facts were being collected and the creation of a really great theory was anticipated in the very near future. This theory was created in the 1920's, and then the nature of atoms and molecules was completely understood. Likewise, nowadays there is also an enormous collection of facts being assembled, which are necessary for the creation of a theory for elementary particles and nuclear forces.

Whenever I meet the man who was my physics teacher in school, I always express my respect to him. I cannot explain the inner timidity that I feel toward this man, even though I have long known that my teacher was always rather weak as a physicist.

I became seriously interested in nuclear physics after reading the book *The Attack on the Atomic Nucleus* by A. Wal'ter, which was published in 1934 or 1935. Many years later, when I had already become a member of the Academy of Sciences and director of the Institute, I met the author. For a long time he was unable to accept my attitude toward him, the attitude of a pupil toward his teacher. Up until his death I treated him with deep admiration, this man whose words had once made me understand the beauty of a wonderful world.

A pupil must have respect for his teacher all his life, even if he has considerably surpassed his teacher in terms of scientific accomplishments. It is not unacceptable for a pupil, after having grown up, to feel even antipathy for his teacher (because, after all, there are many different kinds of teachers); but, in accordance with the ethic that has taken shape over many centuries, he has no right to go against his teacher. He is able to leave him — and this is the extreme form of protest by a pupil.

My thinking is that a pupil must never go against his teacher. It is almost impossible to imagine a case in which such behavior could be excusable. A pupil may express an opinion of his own that, for example, contradicts his teacher's opinion, but he has no ethical right to contend with his teacher. I know that this assertion, no doubt, is disputable. Nevertheless, let us continue with our line of reasoning.

The laws in our lives that turn out to be correct are not only the laws of logic, but also the laws of morality, the violation of which lead to the degradation and ruin of entire societies. If you are not in agreement with your father, you can leave his house, but you have no right to struggle against him in his house. This "ironclad" maxim ("honor your father") has existed in all ages and in all countries. I think that those countries that neglected this moral became weakened and eventually perished. In these societies, interest in children and their upbringing and in imparting to them the experience of life were probably lost, and naturally, in the absence of such experience, the children necessarily had to suffer many hardships and perish in the struggle.

If this dictum is rejected in the world of science, then either scientists will no longer try to have pupils or they will take as assistants those with no promise, who due to their weakness are unable to go against the teacher. Even the instinct to reproduce in science is a consequence of this dictum.

The fact that my assertions are so categorical does not at all mean that the law of not daring to struggle against one's teacher is an absolute law. My purpose is to attract attention to this issue. The pupil is neither slave nor servant of the teacher, not even his subordinate. He is his son, along with all the ramifications of this fact, including the usual problems between fathers and children.[2]

As the director of the institute where for the past ten years the youngest academicians in the country have worked and where the average age of the members of the Scientific Council is about 30 (I won't even mention the fairly common situation we have in which people who are 25 to 30 years old defend their doctoral theses), I can express some judgments without being afraid of being called an oppressor of youth.

There exists an official definition of "young scientist," and there are even competitions involving papers by young scientists. Officially, this category includes persons under the age of 33. We have no idea where this "age of Christ" came from. I will only say that no such category exists in our institute. Was it appropriate to call Sasha Skrinsky, the leader of the largest laboratory within our institute, a "young scientist" half a year before he was elected academician? Or to give this title to Mitya Ryutov, who leads the largest fusion laboratory, just because he is 30 years old? Even Volodya Balakin, who was recently awarded the Lenin Komsomol prize, felt hurt when someone referred to him as a "young scientist" because he was 25 years old.

Isn't it insulting for scientists to be distinguished into separate "weight" categories on the basis of their ages, all the more so because many of the greatest scientific discoveries have been made by persons under the age of 30 years old. One can speak about competition among students and post-graduate students, but not among young scientists. I think that the notion of "young scientist" is appropriate only in an ethical sense: those who are younger must always respect those who are older and understand that experience is not the most unimportant thing in science.

In organizations whose leaders have a correct attitude toward young people, there exist no problems like those of fathers and children, and, in essence, there is no category of "young scientist." In those scientific institutions where conflicts between generations occurs, however, it is necessary, in my opinion, to search for reasons deeper than merely the age differences of the scientists.

What can I say about my own pupils?

[2]Trans. note: The phrase "fathers and children" is an allusion to the title of the famous novel by Turgenyev.

My first pupil is now a professor at Dubna, V. Dmitrievsky, a man two meters in height. I can hardly talk about all the others: somebody will feel hurt that I forgot to mention him, and somebody else will regard it as over-assurance in oneself. It is always easier to speak about your teacher: in this case you yourself can judge. Anyway, from among those who came to me as young beginners, usually as students, and who worked in direct contact with me, no less than 30 of them have become doctors of science and leaders of laboratories and institutes. Fundamentally they are good scientists and persons of high ethical principles. Therefore I have some right to say that our scientific family — by which I mean the scientific staff of the Institute of Nuclear Physics of the Siberian Division of the U.S.S.R. Academy of Sciences and those who had their "origin" there — is sufficiently robust and "healthy." Even those of our children with whom I am not very satisfied (such things happen in every family) are persons of acknowledged high professional level. Perhaps we have some higher requirements?

I would like to refute the common opinion that people move to Siberia for high titles, because I am able to make a judgment about this with our Institute as a case in point. People come to us for enhanced opportunities for work, not for titles. It has even been noted that if someone leaves us to go to Moscow, Leningrad, Kiev, or other cities, he is usually offered a higher position.

From our Institute have come the director of the Institute of Space Research of the U.S.S.R. Academy of Sciences, academician R. Sagdeev; the rector of Novosibirsk University, academician S. Belyaev; the director of the Institute of Automation of the Siberian Division of the U.S.S.R. Academy of Sciences, Yu. Nesterikhin; the deputy director of the Institute of High-Energy Physics and leader of the Serpukhov accelerator facility, corresponding member of the U.S.S.R. Academy of Sciences A. Naumov; and many others.

In our work of training scientists, were there defective products? Yes, of course, just as happens in any kind of manufacturing. However, their number is insignificant, if I may use this word in reference to people. All the more I am very happy with our young laboratory leaders who have grown up within our Institute, usually from students doing their training work: among them are academician A. Skrinsky, corresponding members of the U.S.S.R. Academy of Sciences V. Sidorov and L. Barkov, professors D. Ryutov and V. Volosov, as well as many others.

Now I would like to return to the subject from which I began — the necessity of the development of real scientific schools. Experience has shown that science, just as life itself, cannot give birth to itself out of nothing. Regardless of the financial resources that may be expended, they do not by themselves cause scientific research to arise on a really scientific scale. The fruitful development of scientific ideas is hardly possible where there are no primary scientists and scientific school. Using only financial resources, one can create only the illusion of science, but not, as a rule, real science. There are hundreds of examples of enormous institutes being built and then stuffed with first-class equipment and provided with extensive funds for paying the employees, but where there is no science, perhaps only the outward appearance of science. Here, however, is an opposite example. The small country of Denmark is obviously unable to invest enormous resources in every kind of research, but nevertheless it occupies a leading position in the field of the physics of the atomic nucleus. This occurred thanks to the famous school of physics that was established by the great Neils Bohr. This is precisely a school that determines the level of scientific work and on which depend the forming and preparing of researchers.

There is no scientist in our country who does not know the school of academician Ioffe at the Leningrad Physico-Technical Institute. This was the school that became the foundation for almost all physics research in our country. It produced such famous scientists as Kurchatov, Alikhanov, Skobeltsyn, Alexandrov, Artsimovich, Semyonov, and many others, whose names are already associated with their own scientific schools.

Likewise, our Institute did not appear out of nowhere. It was organized out of the Laboratory of New Acceleration Methods that I headed up, which had existed at the Kurchatov Institute. By the time when the Siberian Division of the Academy of Sciences was created, it was a large laboratory with established topics and scientific traditions: 140 people and several trainloads of equipment moved to Novosibirsk.

I myself graduated from the university in 1941 and immediately went into the army and was sent to the front. After being demobilized following the war, I began to work under Kurchatov. When I recall the initial years of working on the nuclear problem in the U.S.S.R., to me it seems that it was not science, but poetry. Music, even! The very activities of the people who were occupied with this seemingly difficult and, in the opinion of many nonprofessionals, fearful work can be characterized as poetic. They created a symphony of happiness and beauty. In terms of beauty and grace, no equation was inferior to a Venetian vase.

Nowadays people are accustomed to hearing conversations about atoms and nuclei, and so much progress has been made in this field that many things now seem primitive and naive to us. But let us go back something like 25 years ago. We can see what an explosion of human knowledge occurred then — such wonderful things, such great discoveries, not only scientific ones but also discoveries of social significance for all people, to which everything that came before in the history of science cannot be compared.

Nowadays we can hardly imagine how important at the time was each new pronouncement and each discovery, however small, along the way as progress was made toward the final objective. I remember these three years of working every day until two o'clock in the morning, without any days off or vacation, as the brightest and most delightful years of my life. Never thereafter have I listened to music or read poetry, nor in general can I conceive of any work of art, that can be compared to the activity of solving the nuclear problem, in terms of the beauty of its inner essence and outward form and its harmony of sense and reason. In those days it was difficult for us to imagine that there could be a symphony that could sound better than the music of experimental results.

Of course, this beauty, harmony, and elegance are not accessible to everybody. For example, many of the perfections in music also remain inaccessible to the uninitiated. Beethoven's music was not understandable to many of the well-known experts of his day. Many people are still unable to comprehend the beauty of modern sculpture. Therefore, in order to be able to discern beauty and elegance in something, it is clearly necessary to be trained to some extent. People who do not know quantum mechanics and the theory of relativity are unable to perceive all of the beauty and elegance of modern physics. Unfortunately, on account of this, they lose out on a lot. The fact that they do not perceive this beauty, for which nobody blames them, does not, however, give them the right to assert that it does not exist at all. But those who once come into contact with these, the supreme creations of human reason, will discern, in addition to deep thought, also beauty — emotional beauty that has the same effect on people's feelings as music, poetry, and painting do. An expert sees and feels all of this. And I invite you young people to have this feeling. For this it is necessary to study, and to study a lot.

Everyone who chooses science as his future embarks on a long voyage. Whether you will land on fair and abundant coasts or on hostile and barren ones depends on many things — first and foremost on yourself.

I have mentioned that it is the custom to wish fair winds to everyone who embarks on a long voyage. But if the ship has a strong rudder and an experienced helmsman, it can travel not only with the wind, but also across it and even against the wind. Moreover, if the wind always blows on your back, stop and reflect: Are you going in the right direction? Aren't you just going where the wind wills? It is very dangerous to follow the wind in science: you constantly have the illusion that you are moving, whereas in fact you are being moved.

The most dangerous thing for a ship is a calm. In this situation it is possible to move only with the help of a tugboat. Therefore you should only be afraid of the calm. You should not be scared of side winds and opposing winds, for with them one is able to move forward toward the goal. Beware of calms!

Get Going!

G.I. Budker

I feel that the changes which have taken place in science in the last few years open up new possibilities, about which I should like to say a few words. In 1951 we began work on thermonuclear reactions in the confident belief that we would solve the problem with a rush and immediately. I was assigned the task of ensuring that our future thermonuclear reactor would not get too much out of hand. It was like the story of the man who wished to invent a perpetual motion machine and had taken out a patent on a method for keeping it under control. This attitude stemmed from the successes in developing "explosive thermonuclear reactors," a task which was achieved within a

Originally published in *Plasma Physics and Controlled Nuclear Fusion Research*, Proceedings of the 3rd International Conference, Novosibirsk, 1-7 August 1968 (International Atomic Energy Agency, Vienna, 1969), Vol. 1, pp. 41-42.

very short period of time, leaving physicists with the impression that they could do everything — and do it fast. However, experience soon showed that here we had a scientific rather than a technological problem and that it would be necessary to study in detail the physics of plasmas — which we have now been doing for over ten years.

The work in this field attracted new people with a new philosophy and a new ideology. Now, I feel that the progress achieved by the physicists during this period again justified our thinking in terms of building a thermonuclear reactor. The physicist is not obliged to embark upon a project only when he is in possession of all the facts; he does not have to wait until the last button is sewn onto the tunic of the last soldier before engaging battle. He needs only to study the underlying principles carefully and then to find a solution which will reveal the unknown. Such was the situation in the case of the first atomic (uranium-graphite) reactor.

That does not mean that we should give up plasma research — quite the contrary in fact; it should be remembered that nuclear physics really got under way only after the first atomic reactors had been built. However, I feel that the amount of data accumulated is now sufficient for some of those working in the field of plasma physics to direct their attention to the construction of a thermonuclear reactor, leaving the research to those who prefer to study physics problems. If it is objected that no new ideas have been advanced as to how this should be done, my answer would be that ideas materialize in the course of work and that, if we do not change our philosophy, we shall resemble the sophist who said that one should not enter the water until one had learned to swim.

The question is often asked as to how long it will take to build a thermonuclear reactor. The answer to this question is reminiscent of the story about the wise man who was asked by a traveller how long it would take him to reach a town. The wise man replied, "Walk on! Walk on!". The traveller was nonplussed and shrugged his shoulders, but the wise man said, "Walk on, and then I shall tell you." The traveller walked a little way and then turned around and looked at the wise man, who said, "Keep walking, don't look around." So the traveller continued walking straight ahead. At length the wise man said, "Now I can tell you how long it will take you to reach your destination — now I know how fast you walk." When we have considered the problem as a whole and ascertained what material and human resources are being expended, we shall be able to answer the question how long it would take to build a thermonuclear reactor.

There is another question: Who will first reach the goal? Those who first take the road towards the practical realization of our objective will be the first to reach it. One thing is certain, whoever reaches the goal will do so by virtue of the work of a great many scientists and statesmen, the work of the people concerned with this series of conferences. The results of these labors will be the fruit of international scientific cooperation between countries in all continents.

The problem of the thermonuclear reaction is an unusual problem of physics, and one which will transform human society and the world. Our generation, which gave the world atomic energy and thermonuclear energy in explosive form, is now responsible towards Mankind to solve the main problem — obtaining energy from water. The world expects it of us, and it is our duty towards mankind. It is a task which our generation must accomplish, and to do so we must now set forth on the road.

"Engineer of Relativity"
The Moscow Period (1945–1957)

G. I. Budker (center) and I. V. Kurchatov.

At the Kurchatov Institute

I.N. Golovin

The results of Andrei Mikhailovich's scientific activities are completely presented in his collected works, published in 1982, and therefore I would like to devote a few pages merely to reminisce about our meetings.

One sunny summer morning in 1945 while in my laboratory, I was telephoned by Pavel Vasilievich Khudyakov, who at that time was one of the deputies for I.V. Kurchatov at Laboratory No. 2 of the U.S.S.R. Academy of Sciences.[1] He told me: "I gave your telephone number to a young man. Please have a talk with him. Probably he will do for work — he's a good person."

Pavel Vasilievich had keen intuition. He discerned people very well. Soon, the expected telephone call came. A young voice told me about his recent conversation with Khudyakov, and we arranged to meet at dinner time at my home.

---

[1]Editors' note: In the 1940's this was the name of the I.V. Kurchatov Institute of Atomic Energy.

At that time I was unmarried and lived in what was the main building of Laboratory No. 2, in which were located the principal experimental facilities and the administration. At the hour agreed upon, a young man came, dressed in an army blouse, who introduced himself as Andrei Budker.[2] We hit it off immediately. Budker was animated and talkative. After a little conversation, I suggested that he either solve or recall the solution for the betatron problem concerning the radial profile of the magnetic field that is required for an accelerated electron to remain in equilibrium on its orbit. I thought that this question would be difficult for someone who had been in the army all during the war, but I was pleasantly surprised when Budker, who readily began to solve this problem, gave me the correct answer after a short calculation and then immediately went on with a further discussion of the orbit stability condition. We had a lively chat for about another half-hour and then parted on friendly terms. I was absolutely certain that it was necessary to recommend to Pavel Vasilievich to bring him on the staff without hesitation. He (as I learned later) was utterly exhausted from our conversation because he had been hungry and not feeling very well and because he thought his answers to me had been incoherent.

Even so brief a contact with him was sufficient for me to understand that Andrei would find nothing to do in our laboratory, which, under the leadership of L.A. Artsimovich,[3] had begun working on the electromagnetic method for isotope separation. He was too talented for our narrow field. Hence, I recommended him to V.S. Fursov,[4] who at that time was occupied with two problems: theory for the oscillations of ions in cyclotron orbits, a cyclotron having just been constructed, and initial problems of controlling a nuclear reactor (or, as it was called then, a uranium-graphite cauldron).

I did not meet Budker again until the end of 1950. Later I learned that his work in both areas had been very productive. Not only did he participate in analyzing the results from the 1.5-meter cyclotron that

[2] Editors' note: This appointment took place before Budker was sent to the Far East Region. He was brought on staff at the Institute after the demobilization — in May, 1946.

[3] Editors' note: L.A. Artsimovich (1903-1973), academician, was the leader of thermonuclear research in the U.S.S.R.

[4] Editors' note: V.S. Fursov, Doctor of Physics and Mathematical Science, was a three-time recipient of the U.S.S.R. State Prize and actively participated in work on the nuclear problem.

was set in operation at Laboratory No. 2, but he also worked on the large synchrocyclotron project in Dubna. He was also in close contact with the group of N.A. Dollezhal'[5] and, in collaboration with them, developed the control theory for industrial nuclear reactors, which were being constructed in those days.

At the end of 1950, Igor Vasilievich Kurchatov enlisted A.M. Budker, along with other theorists, to work on the new problem of controlled thermonuclear reactions that excited all of us. Here, Andrei Mikhailovich's talents became widely apparent. Ideas gushed forth from him, and he developed them successfully. As he himself acknowledged at that time, he felt wonderful and his work was very fruitful, in spite of some circumstances that complicated his life. A consequence of his investigation of the problem of a toroidal magnetic thermonuclear reactor (MTR) was the birth of his idea for the stabilized relativistic electron beam. Hereafter he worked successfully on two "fronts," continuing to be one of the most active participants in the thermonuclear program, but also developing concepts for new particle acceleration methods.

At that time I was first deputy for I.V. Kurchatov. One day Andrei Mikhailovich asked me to place under his authority a small group of engineers and workers, to work on extracting proton beams from the Dubna synchrocyclotron. He kept persuading me that he had not only calculated everything, but that he also knew how to design and build a channel for extracting 90% of the accelerated protons, instead of the few percent then being extracted. I did not believe Andrei Mikhailovich and so did not assign anyone to him. I considered him to be only a talented theorist, not discerning his ability as an "engineer of relativity" (as L.D. Landau later called him) nor his extremely exceptional talent for organizing collective scientific research! Thus, Andrei Mikhailovich did not obtain from me support for his work at Dubna. However, maybe it turned out for the better, since there he would have worked on solving a particular problem, whereas here, at Laboratory No. 2, he broadly extended the horizons of the field of thermonuclear fusion, as well as the field of new acceleration methods.

One day, when he was already working on the concept of a magnetic-mirror confinement trap, Andrei Mikhailovich had a conversation with

[5]Editors' note: N.A. Dollezhal', academician, originated the first industrial reactor for the world's first atomic electrical power station and also reactors for subsequent electrical power stations.

L.A. Artsimovich, to whom P.E. Spivak[6] had come to discuss the possibility of focusing beta electrons with a magnetic field. What caught his attention was the question of how the electron flow from the target to the counter would increase if the target were to be put in a weak-field region, with the counter in a strong-field region, and if the spiral motion of the electrons along the magnetic field lines were to be taken into account. Budker was astonished to hear that Artsimovich understood perfectly well that the electrons would be reflected from the strong field region for certain values of the angle between the direction of the field line and that of the electron velocity. Artsimovich persuaded Spivak that this method would not increase the solid angle within which the electrons coming out of the target would reach the counter. At that time Budker had not yet finished the solution for the Fokker-Planck equation and had not told anyone about his trap with two magnetic mirrors ("plugs"). As he sat in Artsimovich's office and listened to him fervently convince Spivak that the electrons would invariably be reflected by a strong magnetic field, Andrei Mikhailovich waited, trembling, fearing that Artsimovich would make one more little step in the reasoning and come up with the "Plugatron" idea. But it did not happen! Relieved, Andrei Mikhailovich left Artsimovich's office and began to calculate intensively.

At the end of 1954 he described his ideas to the thermonuclear scientists of our institute, thus founding a new direction in controlled thermonuclear fusion, that of open magnetic traps. His related work was included in the four-volume *Plasma Physics and the Problem of Controlled Thermonuclear Reactions*, which was published through the initiative of I.V. Kurchatov prior to the Second Conference on the Peaceful Uses of Atomic Energy, held in Geneva in 1958.

In science, walking on the brink of discovery like this is not rare. We who were involved in the fusion project witnessed another no less striking example. In the 1960's, the American physicists Fowler and Rankin had shown that a positive plasma potential in an open trap, together with the slowing down of ions on electrons, led to a significant reduction of the ion confinement time in the trap. Kelly then proposed that both ends of the main trap, in which the thermonuclear energy must be generated, each be joined with a small plasma trap, in order

---

[6]Editors' note: P.E. Spivak was a nuclear physicist and a corresponding member of the U.S.S.R. Academy of Sciences.

to eliminate the electric field-induced pushing of ions out of the main trap along magnetic field lines. Physicists around the world read this proposal and discussed it, but it took almost ten years before G.I. Dimov and his collaborators in 1976, and subsequently Fowler and Logan in the United States, took the next step — in essence a very small step — to the qualitatively new concept of ambipolar confinement of a plasma in an open trap. What is required is that the plasma density in the trap joined to the main trap not be equal to the plasma density in the main trap (as Kelly had done), but instead higher.

Budker's report in 1954 about a new method for plasma confinement received a guarded response from theorists, including academician M.A. Leontovich. It took one year to prepare the first experiment, performed by M.S. Ioffe[7] on the PR-1 device. And, it was only in 1957, after three years had passed, that the OGRA project was begun. This latter project was the realization of Budker's idea, formulated in his fundamental work of 1954, about creating a plasma by molecular ion injection.

The idea of a stabilized electron beam also did not find much support in the first year after its publication. It was D.V. Efremov[8] who helped by persuading I.V. Kurchatov to create without delay a laboratory for new acceleration methods under Budker's leadership. The first step in this work was the involving of the talented engineer and high-frequency specialist A.A. Naumov, who had put the large cyclotron "on its feet" back in the 1940's. The experiments indicated that the creation of a stabilized electron beam would be a lengthy and difficult task, with no guarantee of success. However, it was vital that the new laboratory demonstrate its capabilities. I advised Andrei Mikhailovich to choose a more near-term goal. Thus, he together with A.A. Naumov began to construct a small pulsed ironless synchrotron, absolutely original in its design. Talented young people built it in a few months. Electron pulses with energy 2 MeV were soon being obtained in a normal laboratory room. Shortly thereafter their energy was increased to 70 MeV. This

[7] Editors' note: M.S. Ioffe, Doctor of Physics and Mathematical Sciences, is well-known due to his work in plasma physics.

[8] Editors' note: D.V. Efremov (1900-1960) was the Minister of U.S.S.R. Electrical Industry (1951-1953) and first deputy to the chairman of the U.S.S.R. State Committee on Atomic Energy (1954-1960).

work was a complete success. The reputation of the laboratory was becoming firmly established.

Andrei Mikhailovich was skillful at enlisting young scientists with talent and enthusiasm. He worked hard, teaching them physics and how to do daring experiments. The scope of research broadened out. Soon the construction of the world's first colliding electron beam device was started. A small one-story building was placed at the disposal of the laboratory. Nevertheless the general attitude toward A.M. Budker at the Institute of Atomic Energy was still one of skepticism. His projects were too extraordinary. Apparently representatives of the older generation, as well as persons the same age as Budker, simply envied him, being inferior in creativity and in boldness of practical decisions.

In 1958 the famous physicist M. Oliphant, who had collaborated with E. Rutherford and who had discovered tritium and helium-3 and — most important for the thermonuclear problem — their nuclear transformations with deuterium, came from Australia to visit the Institute of Atomic Energy. When he visited Budker's laboratory, he was so delighted with the abundance of ideas under development and the originality with which they were being realized that he refused to visit other laboratories at the Institute of Atomic Energy and spent the whole day in the Laboratory for New Acceleration Methods.

I.V. Kurchatov watched Budker's work attentively and saw that it would be impossible for him to accomplish all his plans in the Institute of Atomic Energy. Therefore, during the last two years of his life, Kurchatov very vigorously supported the creation of the Institute of Nuclear Physics in Akademgorodok, in the vicinity of Novosibirsk. Many times, after visits by Budker, I found Kurchatov in good spirits, repeatedly exclaiming, "Good for 'Sir Director'!" (This was what Kurchatov, who always gave nicknames to all his employees, called Andrei Mikhailovich after it had been decided to appoint him to the position of director of the new institute that was being built.) "What a brain! He is developing such a great thing. We must help him." Indeed, he helped him greatly. Thanks to Kurchatov's help, the Institute of Nuclear Physics of the Siberian Division of the U.S.S.R. Academy of Sciences already at the time of its creation had acquired great potential, which allowed it to occupy a leading international role for research in the area of colliding beams and in the thermonuclear fusion program.

A Divinely Favored Physicist

A.B. Migdal

In 1946, after demobilization, Andrei Mikhailovich called me and asked me to put him on the staff. We met at my home. At first I asked him several scientific questions, and it turned out that, after having been in the army, he knew — or, more precisely, he remembered — very little. However, his lack of knowledge had an unconventional form. In response to the question of what is the deuteron spin, he answered, "But that is quite obvious: either zero or one." And, of course, such an answer is much better than saying "I don't know" or having the correct answer memorized. I began to inquire of him what he had been doing in the army. It turned out that he had made several inventions, which had been used in the anti-aircraft unit where he had served. It became clear that it was necessary to bring this person on staff.

I am proud that I successfully saw in this provincial young man (at the time Andrei was 28 years old) the extraordinary character of his thought and the vigor and breadth of his views. Vigor was one of his main features, and it was manifested already at that time in everything he did. He did not standardize his unusual name Gersh Itskovich to, for example, Grigory Isaakovich, but instead, with élan, renamed himself Andrei Mikhailovich.

My department at the Institute of Atomic Energy, which Andrei joined, was a collection of people who were very close in spirit and who were absolutely devoted to science. The environment was extremely active and creative. Andrei adopted the department style immediately, so that with his appearance, our scientific arguments became even more heated. One day we were discussing a very difficult and intricate question, which required our complete attention. With his hyperactivity, Andrei made it altogether impossible for us to concentrate on the question. After several "final" warnings, I just pushed him out and closed the door. Even then, however, he did not calm down, but shouted through the door, "Make the substitution $1/x$!" As the other participants used to tell the story, I held my head in my hands and groaned, "My God, what am I to do?" Although I do not remember, it is very possible that Andrei's suggestion was found to be correct.

There was another time, when our work required doing contour integrals. Andrei declared one day, "I understand the case when a woman has three breasts." He meant that he had considered the case when the contour passes among three equidistant peaks.

One morning after he had arrived at work, Andrei solemnly exclaimed, "I did not sleep last night, but as a consequence I figured out how the ideal collective farm should be organized. To begin with, the collective farm acquires two tractors." Victor Mikhailovich Galitsky,[1] who had precise knowledge about everything, declared, "A collective farm has no right to buy tractors." Andrei responded, "Well, then, my project fails — without tractors, there is nothing to discuss." We never did learn what constituted his scheme.

With youthful generosity, we discussed in the same spirit concrete questions about nuclear physics, as well as fantastic or abstract problems like machines that reproduce themselves. I recall how Andrei and I "made up" the theory for a heterogeneous nuclear reactor. This was theoretical physics, which did not require those qualities in which Andrei particularly excelled. Nevertheless, to be occupied with doing "standard" theoretical physics with him was very pleasant and productive. The most amazing thing about him, the characteristic that put him

---

[1] Editors' note: V.M. Galitsky (1924-1981), a theoretical physicist, was a corresponding member of the U.S.S.R. Academy of Sciences from 1976.

head and shoulders above everyone else in his field, was his boundless engineering-physics imagination. We theorists were far removed from technical matters and had little talent for engineering, but Andrei had it in the highest degree. His projects seemed fantastic to us at first, but later, to our great astonishment, they did become alive with practical reality. This was one of his most wonderful features — the ability to dream, but in such a way that the dream could be carried out.

When all of us accepted I.V. Kurchatov's invitation and began investigating the problem of controlled thermonuclear reactions, Andrei was the one who almost every day came up with new ideas for us to discuss. During this period, his extraordinary talent for physics inventiveness matured. I remember when he came in one morning after a sleepless night and described to us the concept of the magnetic trap, which he had just devised. The term "physics inventiveness" should be specified. Andrei was never content with only an idea. He examined a problem from all angles, using whatever was necessary from the arsenal of theoretical physics — figuring the integrals of motion, solving the kinetic equation, checking stability. An all-round theoretical investigation of engineering-physics ideas was the most significant characteristic of his major works.

One more extraordinary thing: He had a complete knowledge of technical capabilities, even though he learned about technical matters, just as he learned about physics, not from books but "by listening." Without this knowledge, his unfettered imagination could have led him far astray from reality.

Our personal relationship transformed from a close friendship into a mutual affection. This affection did not weaken after Andrei moved to the Novosibirsk scientific town. Our friendship did not prevent us from joking with each other. At the banquet after the international conference held in Novosibirsk Academgorodok, I told a story about two parrots: "One parrot could sing and dance and play the guitar, but was priced cheaply, whereas another parrot could do nothing, but cost twice as much. When a customer asked why the second one was more expensive, the salesman answered that the second parrot was the artistic director of the first one." I then concluded the story with these words: "I propose a toast to the artistic directing of the Institute." Andrei responded, "I accept this toast, but I must mention that I learned the art of directing people during my ten years under Professor Migdal."

Andrei was already an academician by the time I was being elected to the Academy of Sciences. During the discussion that preceded the elections, he said to Igor Evgen'evich Tamm,[2] "You are applying your efforts to elect a student of yours, whereas I want my teacher to be elected."

I have always thought that when a person grows older, he becomes more experienced and more careful, but not wiser. This adage did not apply to Andrei, for as he grew older, he became wiser, even acquiring the appearance of a Biblical sage.

Once Andrei began a lecture with the statement, "Physics is such a fair lady." This "fair lady" favored him. He was a born physicist — not an experimentalist, not a theorist, but a Physicist with a capital letter.

It was my pleasure to have known him closely and to have learned many things from this wonderful person and great physicist. It makes me proud that he considered himself to be my pupil.

---

[2]Editors' note: I.E. Tamm (1895-1971) was an academician, Hero of Socialist Labor, and winner of the Nobel Prize and the U.S.S.R. State Prize.

Ideas in the Billions

V.I. Kogan

For 30 years, beginning in 1947, I was an acquaintance and friend of Andrei Mikhailovich Budker. During these 30 years, however, we had a close association only during the first 10 to 12 years, in the "Kurchatov" period and partly in the transitional "Moscow-Siberian" period of his life; in the later years, our contacts (usually in Moscow, seldom in Novosibirsk) were only sporadic. My memories are structured and colored accordingly. May those people who mostly knew the later "Novosibirsk" Budker not be in a hurry to attribute possible differences in my perception of his image to my "aberration." More likely, the real reason is that Budker himself changed gradually as time passed.

1. "A Young Staffer" (Gipokonin, Fusion, and Thereafter)

"So what about Budker, your young staffer?" This question was asked with interest by Igor Vassilievich Kurchatov to the head of the theoretical section, A.B. Migdal, when he met him near the main building of the Institute of Atomic Energy sometime in the early part of

1951; the section was then working on the problem of controlled thermonuclear reactions.

"The young staffer" by this time had worked almost five years at the Institute; had acquired an outstanding reputation in at least two lines of applied research, nuclear reactors and accelerator physics; and had been honored with the Stalin Prize[1] (1951). Just before this new milestone along his scientific path, he had been successfully involved with basic nuclear physics[2] and had actively collaborated with V.M. Galitsky,[3] A.A. Kolomensky,[4] and M.S. Rabinovich[5] in developing the theory of the "cyclosynchrotron" accelerator.

And so Budker dove head-first into the development of the subject of magnetic thermonuclear reactors (MTR). He put forward some pioneering ideas in the physics of hot plasmas. In particular, he was the author of the first estimates of the effect of run-away (or, in his own terminology, "whistling off") electrons; of the analysis of transport phenomena in toroidal plasma systems (or the theory for the "mixing" of trajectories, initiated by academician I.E. Tamm, which later grew into neoclassical transport theory); of the "betatron" method for heating a plasma; of an independent rediscovery of the non-diffusive nature of resonant radiation transfer (the Biberman-Holstein equation); of the theory for multiquantum recombination in a plasma; and of the kinetic theory of relativistic plasmas (the last works were written together with S.T. Belyaev).

In 1953, A.M. proposed a fundamentally new type of thermonuclear trap — a system with magnetic mirrors ("plugs"), which soon became the subject of widespread elaboration both in the Soviet Union

[1] Trans. note: At that time, this was the name for what is now called the U.S.S.R. State Prize.

[2] Author's note: Contour integrals played such a great role in these investigations that we temporally renamed our group Gipokonin, from the Russian acronym for "State Institute of Contour Integrals." [This pompous-sounding name provokes some irony, since "konin" sounds like the Russian word for "horsemeat" (which was eaten in times of hunger).]

[3] Editors' note: V.M. Galitsky (1924-1981) was a corresponding member of the U.S.S.R. Academy of Sciences. At that time he was a young theoretical physicist in A.B. Migdal's group.

[4] Editors' note: A.A. Kolomensky, professor, specializes in charged particle beam physics and accelerator technology.

[5] Editors' note: M.S. Rabinovich (1919-1982) was a professor and a specialist in plasma physics and accelerator technology.

and abroad. In 1952-53, he proposed the so-called stabilized relativistic electron beam, the latest outworking of which in the Institute of Atomic Energy and in the Institute of Nuclear Physics of the Siberian Division of the U.S.S.R. Academy of Sciences stimulated an intensive development of several directions in accelerator physics and technology. It is not superfluous to mention that, precisely at that time, Budker survived a difficult period, having been removed from the project for reasons unknown to him.

Soon it became clear that the framework of only theoretical activity was too narrow for A.M., and in 1954, supported by Kurchatov and Artsimovich, he began directly leading the experimental elaboration of his own ideas.

In 1957, a completely new stage in Budker's life opened up: Igor Vassilievich Kurchatov put him in charge of organizing the Institute of Nuclear Physics of the Siberian Division of the U.S.S.R. Academy of Sciences. For the next 20 years, this institute was permanently headed by A.M., and the work of this prominent scientific center under his leadership went precisely in accordance with the directions that grew up from his own physics ideas.

Now let us go back a number of years.

2. Brushstrokes to the Portrait of the "Early" Budker

A.M. liked to laugh, joke, and tell and listen to anecdotes. In the first years of our acquaintance, his humor had a natural imprint of some provinciality, as well as five years of military service. (I remember the expression of one sergeant major that he quoted with pleasure: "Orderliness — this is when something is lined up with something else.") However, as the years passed by, his humor kept on becoming more sophisticated. For example, in his younger years A.M. joked about a physicist from Dubna who, in A.M.'s words, was regarded as an experimentalist by the theorists, and as a theorist by the experimentalists. Later, sometime in the 1960's, in response to the not very modest reasoning of a Swedish physicist that his country flourished as a result of its always being peaceable, A.M. reminded him jokingly that the thanks of the Swedes for such a blessed state should also be given to our country (an allusion to the fate of Charles XII).[6]

Also, here is a humorous verbal duel between A.M. and R.Z. Sagdeev,[7] which also deserves to be mentioned: "Why did you torment our Russia?" — "And why did you crucify our Christ?"[8]

In his relationships with friends (and even with people he did not know that well), A.M. was kind and attentive. I remember how, at the end of 1947, he, an "elder" of the theoretical section, did a lot of running over to the accounting and personnel departments in an effort to have the first salary to V.S. Kudryavtsev, one of our scientists, and to me be paid more quickly. I also remember how, in the summer of 1951, in spite of his extremely busy work, he took time out to take a long trip to the suburbs in order to congratulate my wife and me on the birth of our daughter.

In A.B. Migdal's office there was a washstand. The stream of water running from the tap deflected noticeably when a comb, electrified on something, was brought near to it. I called this apparatus a "ratiometer" ("ratio" in Latin means "reason"). Budker demonstrated that this effect was much stronger from his trousers (a little below the back) than when I used my hair to electrify the comb. This was an unambiguous indication of our different physics competencies. That's just the way it was....

A.M.'s dealings with language were far from standard. His knowledge of foreign languages was very poor. As for his Russian, he had problems with spelling (perhaps this is why he said many times that correct writing is a typist's job). I remember some examples of his spelling: "mas" (mass), "osocellator," and — probably the record — "inergy," which he wrote on the blackboard during a presentation in

---

[6]Trans. note: The Swedish king Charles XII was defeated in 1709 by Peter the Great of Russia, after which (Budker implies) the Swedes had good reason to become peaceable.

[7]Editors' note: R.Z. Sagdeev, academician, is a theoretical physicist. He worked at the Institute of Nuclear Physics during 1961-1970 and thereafter became the director of the Space Research Institute.

[8]Trans. note: To understand this riposte, it is necessary to bear in mind that Budker was of Jewish ethnic background, whereas Sagdeev was descended from the Tatars.

Dubna (sometime around 1970). (Incidentally, his handwriting was like "chicken scratching.")

However, Budker's oral speech, as well as the style of his writing, was exceptionally smooth and fluent. Therefore, he not only had plenty of reason for dictating his papers directly in the final version, but also enjoyed doing so.

One more very original detail: A.M. often and, moreover, quite involuntarily pronounced words as if they were constructed from correct, but mutually mixed up, syllables or "pieces." For example, *baldakhon* came from combining *baldakhin* ("canopy") and *balakhon* ("loosefitting robe"); *Semiralda's gardens* from *Semiramis* and *Esmeralda; chvakat* from combining *chavkat* ("munch") and *kvakat* ("croak of frogs"); and so forth. One day during a group discussion, he said to V.S. Kudryavtsev, "Vasya, what are you muttering *vtikhomyatku?!*" Vassily Sergeevich was quick to retort: "Andrei, not *vtikhomyatku,* but *vsukhomolku!*"[9]

It seems to me that this sort of interchangeability of syllables was not an accidental defect in A.M.'s oral speech. More probably, it was a natural consequence of a heightened flexibility, of the lability of his entire "second signal system,"[10] which in turn was directly related to the strikingly creative nature of his intellect.

A.M. was very sociable and active in giving various kinds of advice. In particular, this trait was displayed in the following episode. There came to work in our section a young floor polisher who did the work "with his feet" (electric floor polishing machines were still a rarity at that time). The following colorful conversation that took place between him and A.M. requires almost no commentary:

Budker: "You would be better off to invent a machine to polish the floors."

[9]Trans. note: Budker combined the first part of the word *vtikhomolku* ("whisper") and the second part of the word *vsuchomyatku* ("cram dry food into one's mouth") to make up an adverb, *vtichomyatku,* to describe what he thought was muttering that interfered with the discussion. Vassily Sergeevich, in turn, cleverly combined the other parts of the same two words to make up another real-sounding word, *vsuchomolku,* to describe his behavior.

[10]Trans. note: According to I.P. Pavlov's theory for higher nervous activity, the "first signal system" refers to conditioned-reflex nonverbal sensations received from the environment, whereas the "second signal system," which is unique to human beings and which evolves during creative activity, refers to the ability to generalize the primary signals into words.

Floor polisher: "Nooo.... To do that I would have to learn advanced mathematics — the various hyperbolas and preambles[11] and all that stuff. That's too difficult for me...."
Budker: "Well, then, what isn't too difficult for you?"
Floor polisher: "Well, for example, philosophy... along the lines of that in Chapter IV of *A Short Course*."[12]

In 1950 I wrote "on Budker" the following verselets:

To Andrei Budker

A dancer with a Ph.D.[1]
(Or a philosopher of dances?),
An expert in the life advances,
Whereto the gaze does fly of thee?

Your powerful imagination
Attracts you to the space of Asia,
As a collective farm creator[2]
Or a Formosa liberator.[3]

Your vivid mind and intuition
Create ideas in the billions.
They are fantastic, they are brilliant,
And they are right (by definition).

The reproduction of machines,[4]
The lunar-driven monthly cycle,[5]
And lots of other disciplines
To learn by heart (with no title).

A bard in science[6], a ladies' guy,[7]
In essence — modest, yet impudent,[8]
A great bald head since being a student,[9]
And what a talent! Huge and high!

---

[11] Author's note: The word *preamble* (here mistakenly used instead of *parabola*) in those days frequently occurred in the speeches that our Foreign Minister, A.Ya. Vyshinsky, continuously gave at the United Nations.

[12] Author's note: It is well-known that this chapter was personally written by Joseph Stalin.

Author's Notes on the Poem:

[1]A.M. liked to philosophize, and when he studied at the physics department of Moscow State University before the war, he earned money by teaching dancing. (Indeed, he danced very well.)

[2]A.M. dreamed precisely about leadership over a Siberian collective farm, and this dream was realized in the form of the Institute of Nuclear Physics. I don't remember his system for a hypothetical collective farm, but it can be reconstructed in retrospect from the system of the real Institute.

[3]This is the original Portuguese and later the Japanese name for Taiwan. Just at that time (1950) the liberation of Taiwan was a hot topic, and A.M. suggested solving this problem with the use of rafts in the night, using the fact that "all Chinese look alike."

[4]A.M.'s cybernetic invention (see later in this article).

[5]A somewhat smoothened name, which I thought up to indicate Budker's idea about the intimate connection between certain rhythms in the human organism and lunar motion (this connection was motivated by the concrete mechanism of maintaining an expedient order of life among primitive families).

[6]Budker was so called by Migdal, who meant Budker's permanent wish to work only on problems he was interested in.

[7]I inserted this expression merely to support the rhythm and rhyme; at that time it was not yet applicable to A.M.

[8]"In essence" was one of Budker's favorite expressions.

[9]None of us ever saw Budker otherwise.

A.M. had perfect facility in doing calculations. His inexhaustible inventiveness manifested itself in thinking out novel, even purely mathematical artful dodges (a non-standard representation of the delta function, integration with multivalued Riemann surfaces, and many other things). Moreover, his physics intuition allowed him to foresee which particular terms in a sum of many terms must "mutually cancel," and often he accordingly wrote down (in advance!) "plus" and "minus" signs in front of individual terms.

A.M. always noticed when physical laws were violated. Thus, in the wonderful science fiction novel *Plutonia* by the famous geologist V.A. Obruchev, his attention was caught by an indisputable inconsistency —

namely, the violation of Gauss' theorem, which says that the gravity inside a hollow Earth must depend only on the central body (Pluto), and not on the external layers of the Earth itself. Therefore, the travelers would inevitably fall onto Pluto!

Budker was a naturally gifted person — it is even possible to say that he had no genealogy. The name "Budker" itself, when translated from Yiddish, means nothing more than "bath attendant." A.M. grew up without his father, who had been killed by soldiers of Petlyura.[13] His mother was a woman of little education.

A.M. had a keen realization of the cost of bureaucracy. Many times he said that if all the bureaucrats were dismissed from the civil service, even if they were moved to the beach of the Black Sea and allowed to keep all of their government subsidies, just this change alone would result in innumerable benefits for the national economy and for science.

A.M. was proud of the fact that he came from the same locale (near the Ukrainian town of Vinnitsa) as the famous surgeon N.I. Pirogov. In difficult moments of our country's history, he frequently quoted these words of Pirogov: "We all have only one Russia...."

Naturally, A.M. was a very active and competent participant in scientific seminars. Sometimes, however, he got too "carried away." Once he became "over-enthusiastic" and offended a famous (and, in age, more senior) physicist, declaring to him in front of everyone else something to the effect that "One man is able to ask so many questions that even a hundred clever people will never be able to answer them...."[14] He had another, not very tactful, habit — if he had any doubt concerning the main point of a work that was being presented, he would immediately try to "denounce" it in one swat. Sometimes he succeeded, but much more often he did not — after all, most of our physicists were not born yesterday. Nevertheless, there was still real benefit that resulted from these attacks of Budker.

[13]Trans. note: Petlyura was the leader of the Ukrainian nationalists who fought against the Soviet government during the Civil War after the 1917 Revolution.

[14]Trans. note: More precisely, the well-known proverb states that one *fool* is able to ask more questions than a hundred clever people can answer.

From his youth, A.M. had an interest in philosophy and was even familiar with the writings of Kant and Hegel. I recall how he was able to prove that definitions like "The superstructure is what is above the basis" and "The basis is what is under the superstructure," when combined together, are not just a tautology, but constitute a statement with substance.

Although A.M. was a very original thinker, he was not free from certain stereotypes, especially in his younger years. Some of them he accepted partly because in those days there was a quite understandable overestimation of environmental factors in comparison with genetic factors. Thus, he seriously believed the assertion made in Soviet schools that the nurse Arina Rodionovna had had a significant influence on the development of Pushkin's genius. Also, when we discussed the results to be expected from a well-known hypothetical experiment, viz., arbitrarily selecting children of white and black races, bringing them up from birth in the same conditions, and after ten years or more measuring their I.Q., A.M. confidently tried to prove that the results would be exactly identical or even in favor of the blacks (the rationale for the former requires no explanation; for the latter, he suggested a special mechanism, that black babies are closer to nature and therefore depend less on their parents' guardianship).[15]

On the other hand, A.M. did not give in to the stereotype (which, if we look deeply at its root, was of the same nature!), prevalent among most of us, his friends, that it was somewhat indecent for his mother to sell at market the excess tomatoes that she raised in the kitchen garden at Budker's Finnish-style cottage. Nowadays it is especially evident that he was the one who was right, and not we who rebuked him for this "commercialism." (It should be mentioned that at this time he was the only breadwinner in a family of six.)

[15] Author's note: It is interesting to point out that some incomplete "natural" experiments of this type have nevertheless been conducted. Beginning from 1961, in the Federal Republic of Germany, there were some studies of the children of American occupational army soldiers, blacks and whites. There was no significant difference in the I.Q. of the children of both races. These results were published in the journal *Social Sciences Abroad*, Series 8: Theory of Science, No. 1, pp. 92-98, 1987 (published by INION, U.S.S.R. Academy of Sciences).

The influence of family on A.M., which at that time we also considered to be a "Philistinism", was exhibited in another episode, which had more than one interpretation. We, the Komsomol members in our section, organized among the "scientistics"[16] of our entire department a collection of money for an employee in the administration department, who had become a single mother. The scientists' salaries were high then, and our "big wigs" also responded very generously, so that several thousand rubles (by the old monetary standard) piled up, and we handed it over as it had been intended. Suddenly, however, we came under moral attacks from two different sides. The secretary of the Communist Party bureau of our department, a rather well-known physicist, sternly pointed out to us that such benevolence was inadmissible ("Our state is not poor; she should apply to the local trade union committee"). Also, Budker, sharing the opinion of his wife and mother, laughed at our "naivete," confidently predicting that our gift would by no means be used "for the baby." The stark reality completely confirmed his (or, more precisely, their) prediction: the young woman bought new dresses for herself. However, even today I do not regret this "campaign"....

As already remarked, A.M. initially underestimated genetics. However, he always had a perfect intuition about the other cornerstone of biology, the process of natural selection, as well as the somewhat related matters (which in a sense have a "mathematical-physics" nature) of the "predator-prey" issue and so forth. He transferred the relevant mechanisms to "machine reproduction," wherein machines could absorb their "building materials" from the environment. He thought this up already sometime around 1950, and I suppose he was one of its first inventors.

However, it is interesting to note that he had a kind of "physics egocentrism" that fettered his evaluation (at least in those years) of the prospects of cybernetics: in many ways, he generally attempted to restrict even its conceivable capabilities to only doing calculations (without any heuristic, creative aspect). (His later view on this matter is unknown to me.)

---

[16]Trans. note: A somewhat demeaning term for scientists. (Komsomol was the Communist Union of Young People.)

During his work, A.M. almost never read the physics literature, instead extracting the newest information from numerous verbal discussions. However, I wish to emphasize that he had very deeply assimilated (already before the war) the university physics curriculum — general physics as well as theoretical physics — and, as we could see, this was quite a bit!

Therefore, probably it is not very surprising that A.M. was never able to pass the Candidate examination in classical field theory, even though several times he began to study the appropriate volume by Landau and Lifshitz. Finally, consciously committing a sin, A.B. Migdal gave him credit for this examination (this is exactly that rare case when "adding to the truth" is excused by the fact of the matter!). Indeed, classical field theory was one of Budker's "crowning" subjects, to which he made new and important contributions both then and later.

3. His Place in Physics

Both as a person, but especially as a scientist, Budker presented a very rare and original phenomenon. Here I wish to touch on some of his features as a physicist, the combination of which in one person seems to me to be almost unique.

First, he had the gift of being a "naturalist" in physics (not simply having the well-trained intellect of a theoretical physicist). Related to this, he also had flawless intuition, which allowed him to catch on quickly and understand deeply the mechanisms of physical phenomena, to uncover the main effects, and to evaluate them quantitatively.

Second, with relatively limited scientific erudition (I refer to his acquaintance with the literature), he had firm convictions about the necessity for pioneering research and advanced studies. Many, many times A.M. repeated his guiding principle, to which he resolutely adhered: Not to follow in pursuit of the work of other scientists, but to break his own new "ski trail."

Third, he had inexhaustible imagination and inventiveness, which were the source for many fresh and beautiful ideas in physics. These ideas were based not only on intuition, but also on a thorough knowledge of the capabilities of current technology. Therefore, as a rule, they were exactly in the domains most interesting for science, which are already sufficiently advanced so as not to be trivial, but not yet to the point of being ephemeral. Moreover, he demonstrated the same inventiveness in his scientific administrative activities.

Finally, he had the rare art of combining individual "elementary" ideas into complicated, multicomponent conglomerates, as well as the skill to work them out over months to an impressive level of completion. This feature of Budker's creativity already became clear from his first publications about stabilized beams and magnetic mirrors. Indeed, these were not just ideas, but instead detailed theoretical elaborations of absolutely novel material objects, based on the analysis of a large number of nontrivial physical effects.

L.D. Landau once called Budker a "relativistic engineer."[17] The precise interpretation of this humorous characterization is not known; some people saw in it a note of irony, while others (including the one so called) considered it more likely to be praise. I would like to canonize this latter interpretation. Indeed, how often is it possible to encounter such a scientist, in whom are combined so harmoniously such "polar" talents — and with such irresistible charm?!

This is the person that will remain in the memory of all who knew and loved Andrei Mikhailovich Budker — the pride of Soviet physics.

[17] Author's note: Landau's characterization had much in common (even coinciding in the use of the word "relativistic") with a characterization by V.D. Shafranov, given in verse form, congratulating Budker on his fiftieth birthday (1968): *"The kinetic-relativistic Yermak of physics."* (Yermak was the famous Russian conqueror of Siberia in the sixteenth century.)

Not a Long Life, But a Brilliant One

S.T. Belyaev

Andrei Mikhailovich Budker was a unique figure in our science. An original person of natural gifts, a scientist who really cannot be placed in any particular recognized scientific line or school, he lived a short but brilliant life and left many basic ideas and results.

This is the way, or close to the way, that historians of science have written about Budker and will write about him in the future — and it is all absolutely true. However, for the people who worked with him, who knew him in the whirl of work during the week and in everyday quandries, Andrei Mikhailovich (or simply A.M.) was first and foremost an original and singular personality. Some people were flabbergasted by the unorthodoxy of his motley and multivalued nature. He was the symbiosis of a wise and energetic man with the characteristics of a naive and bustling person from a small Jewish town. He had deep inner conscientiousness and decency, along with the slyness of a conjuror at a fair. But who knows? Perhaps if he had not been such a complicated and "uncombed" person, A.M. could not have had such original and paradoxical mentality. Outstanding talent is often accompanied by an erratic character. And so often it is the case that the latter carries more weight in popular opinion and in every way makes it difficult for the talent itself to be manifested. Therefore, in a gallery of paintings of famous scientists, a smoothened iconographic painting can be both pernicious and inappropriate. A.M. Budker's life is interesting and instructive also from this point of view.

I knew A.M. closely for almost 30 years, from the late 1940's to the time of his death. Many episodes and scenes remain in my memory. I will attempt to recall some of them.

In 1949 I was a fourth-year student at the Moscow Physico-Technical Institute. Most of my time was spent being trained in the theory department under A.B. Migdal at the Laboratory for Measurement Instruments of the U.S.S.R. Academy of Sciences (now the I.V. Kurchatov Institute of Atomic Energy). This department was not large, but had quite an outstanding staff. Among the staff members were B.T. Geilikman,[1] V.M. Galitsky,[2] and A.M. Budker. The solution of important practical problems was combined with fundamental investigations. The atmosphere was very democratic, and the discussions were free and often very animated, revealing individual temperments: the refined intelligence and unfailing correctness of B.T. Geilikman; the same traits, but only when not gotten too excited, of V.M. Galitsky; and the powerful and temperamental self-confidence of A.B. Migdal. The attitude toward A.M. Budker was like that toward a talented but very ill-mannered child. He constantly intruded into discussions, even those concerning problems previously unknown to him, and tried to take over. "Andrei, you would be better off to read up on it by yourself" (this was Geilikman). "Andrei, I'll explain it to you later" (this was Galitsky). "Andrei, you've only heard of this half an hour ago" (this was Migdal); "do you really think you can teach me something about a problem I've been occupied with for many years?" Occasionally the explosion would be stronger. Here is the scene: Migdal busy with calculations, with Budker nearby offering advice, "Make the substitution $1/x$." Finally Migdal explodes in fury, "Andrei, go away!" Budker persists, "Well then, Kadya, make the substitution $1/x$!" Migdal grabs

Andrei with his arms, carries him out the door, and locks it. Budker shouts through the keyhole, "Make the substitution $1/x$!"

Budker also had many ideas beyond physics. Sometimes they were quite exotic, ranging from original principles for organizing collective farming, to a theory for "lunar-menstrual cycle." Yet at the same time he was seriously immersed in solving important concrete problems for reactor and accelerator physics. In general, he was more interested in the practical problems of theoretical physics, although he liked to boast about some of his abstract results. He used to repeat with pride that he had figured out all by himself the topology of the asymptotic features of equations in the complex plane (the Stokes lines).

When it came time to do a thesis for my degree work, Migdal "donated" me to Budker ("he always has practical problems"). At the time, I was attracted more by difficult quantum mechanical problems than by the problem that Budker formulated: "There is something wrong with our cyclotron. The beam losses in the first turns are large. You could probably manage to examine the trajectories of the particles and help the experimentalists." I began to look into this problem, and some ideas developed. A couple of times Budker showed an interest in my work and took a look at my mathematical "beautiful ornaments" (the calculation of the field between two dees, performed with the help of conformal mapping, etc.). Convinced that the thesis would be successful, he then lost all interest and left me completely on my own.

Following my thesis defense and a month's vacation, I returned (in February of 1952) to Migdal's department as a permanent employee and found there a new situation of great urgency: viz., "thermonuclear" research was being developed with all intensity. This research was directed by Lev Andreevich Artsimovich,[3] with the theoretical investigation under the leadership of Mikhail Alexandrovich Leontovich.[4] Migdal's department was also mobilized. We worked hard and with excitement. Enthusiastic seminars by Artsimovich alternated with discussions of theoretical problems with M.A. Leontovich. Moreover, all this work was done in strict secrecy — it seemed that thermonuclear

[3] Editors' note: L.A. Artsimovich (1909-1973) was an academician, Hero of Socialist Labor, recipient of the Lenin Prize, and two-time recipient of the U.S.S.R. State Prize.

[4] Editors' note: M.A. Leontovich (1903-1981), academician and recipient of the Lenin Prize was well-known on account of his works in many fields of theoretical physics. He founded scientific schools in radiophysics and plasma physics.

reactions would become a reality in only a few days. Impelled by youthful fervor, I worked extremely hard, rarely looking around at anything else, until one day M.A. Leontovich and A B. Migdal suggested that I "temporarily work with A.M. Budker, to assist him." Only then did I learn that Budker had been dismissed from the classified thermonuclear project (although he had managed to make some important suggestions about the magnetic confinement of plasmas) and was not allowed to attend seminars or participate in discussions.[5] Left to himself, it was precisely at this time that he came up with the idea of a stabilized electron beam. He probably saw that the only possibility for him somehow to remain at the Institute and maintain conditions for work was to promote this idea successfully. I assume that I.V. Kurchatov supported Budker from this side, but apparently even he was unable to do any more at that time (1952-53). My task was to help Budker substantiate and develop his proposal. He already had enough mature ideas about the general physics, but it was necessary to formulate the precise equations, study their solutions, find the possible instabilities of the beam, and so forth. Periodically, on the instructions of I.V. Kurchatov, the results had to be "run by" the leading experts. I remember the meetings with academicians V.A. Fok, I.E. Tamm, and N.N. Bogoliubov.[6] The notion of a stabilized beam had difficulties in being accepted. More and more objections and questions were raised. Answering them required new calculations. We wrote one fat report after another. We would think that all the difficulties had been put behind us, only to have new calculations be required after the next meeting. The sharpest critic was V.I. Veksler.[7] At that time he was developing ideas for collective accelerators, and Budker, in turn, criticized some of his ideas. The discussions between them were quite pointed and not always polite. Once,

[5] Author's note: Let me remind the reader that this time was near the beginning of the anti-Semitic campaign, which had started with the widely publicized case against the Jewish physicians at the Kremlin medical hospital who were accused of being "killer doctors."

[6] Editors' note: All three of them were academicians, Heroes of Socialist Labor, recipients of the Lenin Prize and the U.S.S.R. State Prize, and authors of fundamental studies in many areas of theoretical and mathematical physics.

[7] Editors' note: V.I. Veksler (1907-1966), academician and recipient of the Lenin Prize and the U.S.S.R. State Prize, authored works in the fields of accelerator physics and technology, high-energy physics, and nuclear physics.

while answering many questions from Veksler, A.M. Budker said something like, "In this case, not even a hundred wise men would be able to give a prompt answer."[8] V.I. Veksler, with his touchiness, took this remark as an insult. It took a long time for Artsimovich to extinguish this scandal. Budker himself, however, did not intentionally desire to insult Veksler. I am sure that he considered this remark to be a joke for relieving stress. The habit of having a ready store of proverbs, jokes, and anecdotes and "serving them up" without any special consideration for their appropriateness remained with A.M. to his last days. His mannerism of telling the same jokes and anecdotes repeatedly was the subject of unceasing teasing. He and his friends took it laughingly, but many people found this style unacceptable. In addition to Veksler, Budker also considered Artsimovich to be a "tough" person to talk with.

Our intensive collaboration continued for more than four years. It was very productive and, I think, useful for both of us. Nevertheless, it did not always go very smoothly. At that time we held different opinions about the criteria for rigor and lucidity and about correlating mathematics and physical intuition in the expounding of the results. Our temperaments were also very different. Our arguments were often loud and animated. Sometimes A.M. would suddenly become quiet and say mournfully, "Spartachishka,[9] why are you, Sir, shouting at me? Am I not your teacher?" Many years later he repeated this remark, but using the intimate word for "you." Sometimes, however, he would retreat: "Together let's write up those matters that don't cause objections, and as for your 'stuff,' do it by yourself and include it in your thesis."

In 1954-55, Budker's situation gradually improved. It was getting to the point that a special department was to be established for the realization of his ideas. However, I.V. Kurchatov did not dare (or was unable) to designate Budker as the leader of this department. In principle, his misgivings were understandable, since in his eyes Budker was

[8]Trans. note: Budker's response was adapted from a well-known proverb, which in its entirety says, "One fool can ask more questions than a hundred wise men can answer." Hence Veksler's reaction.

[9]Author's note: Spartachishka, or "little Spartak," was Budker's familiar nickname for me, constructed from my given name. (The humor in using this diminutive form derives from the fact that my name, Spartak, is etymologically related to that of Spartacus, the Roman gladiator.)

a theorist, who was rather eccentric and who had absolutely no experience in leading an experimental group. At the same time Kurchatov recognized the importance of the work that was being started and made a search, which was not confined within the Institute, for a leader to serve as Budker's assistant. A suitable candidate, in Kurchatov's opinion, was found at the Physico-Technical Institute in Leningrad. But the new leader was someone who had little interest in the main work, and Andrei Mikhailovich was very depressed by the situation. It seems that from this time onward, A.M. had a very harsh and angry opinion of science officials who would set about any project without even understanding anything about it and who expected to "hire" the necessary experts. Fortunately, I.V. Kurchatov had fairly quickly understood the real situation, and A.M. became the head of the department, which turned out to be the seed from which would germinate his pet creation — the Institute of Nuclear Physics in Novosibirsk.

Our paths separated for a while. I remained in Migdal's department as a theorist and became interested in other problems. In 1957 I was dispatched to the Niels Bohr Institute in Denmark. When I returned from Copenhagen at the end of 1958, I learned about Budker's decision to move to Novosibirsk. He was happy and excited and told me about his plans for organizing an institute at this new place, far from Moscow, "where the big scoundrels will not come by themselves, and the little scoundrels can be left behind."

Our contacts were casual until Budker finally moved to Novosibirsk. However, in 1961-62, A.M. began to be increasingly persistent in inviting V.M. Galitsky and me to come to Novosibirsk, "at least as visitors." In the beginning of 1962, we came and had a look and surprisingly quickly decided to move there altogether. Within three months, Galitsky and I were employed at the Novosibirsk Institute of Nuclear Physics.

As a physicist, inventor, organizer, and scientific leader, Budker matured and realized his potential at this Institute. This will be a large and profitable subject for critics and historians of science in the future.

Together with the INP, Budker flourished and grew wiser. In the last years of his life (following a serious illness), he acquired the enlightened wisdom of an older man who has calmly realized the limitations of his time. I remember exactly my last conversation with him, the day before his death. We walked in the small glade in front of our houses (we were neighbors), and A.M. very quietly began to disclose to me his last will and testament: "I feel that the time remaining to me is very short. I think all the time about what will happen to the Institute. I entreat you to give your support to Sasha Skrinsky. I know that you would also be a good director for the INP, and also Sidorov, and also Ryutov. But then it will would be a different Institute. Perhaps a very good one, but different. Also, try to help him." After this, he continued talking quietly about those specific issues, large and small, that we were not to overlook. I tried to encourage him and changed the conversation to lighter topics. But he gently averted my attempts. Apparently, the future of the institute that he had created was the only thing he could think about during his remaining hours.

More than ten years have passed since then. Half a year after his death, I moved back to the I.V. Kurchatov Institute. I became occupied with new problems, scientific as well as administrative. But the more time went by and the more experience I had for myself, the more I became aware how exceptional Budker had been as scientific leader and director of the Novosibirsk INP. I attempted to analyze his unorthodox actions and decisions and to reconstruct his principles.

Although probably unaware of it himself, A.M. had a deep-seated intuitive prejudice against regulations, over-management, and excessive planning. Unceasing creative imagination, both in science and administration — this was his motto. New ideas were constantly being discovered, and he involved the active scientists of the Institute in energetic discussions of them. Sometimes, beginning spontaneously, these brain storms continued for weeks, dying down and then flaring up again. All options might have been considered; nothing might have been found; and the participants in the "storm" might be ready to calm down and return to their regular work. But nothing of the sort! Suddenly A.M. would throw in a new idea, and everything would start all over again. Although most of these "brain storms" were never completed, I think that A.M. valued them highly — to stir up the scientific staff, to prevent any opportunity for minds to be lazy.

Something similar also appeared in the administrative sphere. Budker would constantly propose to make changes in the structure of the Institute, like doing away with one department and creating others. It seemed to me that just as soon as an administrative branch had begun, by normal standards, to run smoothly (with written-up policies and orderly paperwork), A.M. would propose to abolish or reconstruct it. This was his way of killing even the smallest germ of bureaucratization. This was a matter of principle to him. He did not give in even to the firm demands of V.M. Galitsky (the scientific secretary of the Institute), who attempted to establish more order and regulations in the work of the Institute. A.M. survived this conflict with Galitsky (whom he respected and liked very much) with considerable pain. However, he would not give in, even when faced with a threat that Galitsky might move to Moscow (which is what actually happened in 1965).

An extremely important and novel part of the administration that Budker set up was the Scientific Council of the Institute, or the Round Table. It had very little in common with the usual concept of a scientific council. The primary idea was to gather every day at 12 o'clock noon for half an hour to have a cup of coffee and discuss any kind of questions, which, as a rule, were not scheduled beforehand. The discussions differed greatly: serious scientific problems; issues having to do with administration or personnel or housing or international exchanges; talk about a new movie or some local news. Sometimes the discussions would continue for a long time, but uninterested persons could leave freely. From the sidelines, it was possible to think that this was not really very serious. But now I am convinced of Budker's having deeply thought over his idea. For A.M., the Round Table was a place to educate and nurture his associates. Making decisions was secondary. The main purpose was to establish common scientific, moral, and ethical points of view, to teach mutual understanding and constructive overcoming of contradictions and respect for the opinions of others. By himself A.M. could not be the ideal example in everything. Frequently and openly and usually with humor, he discussed many of his shortcomings and foibles. He also very much enjoyed making fun in public of others' mistakes. Although he received criticism of his own actions and decisions without humor or pleasure, he tolerated it and did not keep grudges on account of it. Very often A.M. talked at the Round Table more than all the other people. Nevertheless, his manner of speaking was not that of a leader or mentor, and the atmosphere was highly democratic.

For the active exercise of his mind, to think over and polish ideas, he needed a friendly audience. He was a preacher by nature, and his weapon was a passionate one — the logical persuasion of other participants in a conversation. A.M. felt a creative pleasure in converting die-hard sceptics and even opponents to his own beliefs, always finding those new arguments that were needed exactly at the right moment.

I remember how he once childishly boasted, "At the end of the day I was allowed to have fifteen minutes with the chairman of Gosplan[10] — we talked for two and a half hours. He is a very intelligent and clever person."

But how depressed and lost he felt, when his persuasion had no effect and he had a fiasco! Objectively, his emotional verbosity could be disliked by cool and exact people. Also, of course, it afforded his enemies many opportunities to defame him.

Fortunately, I.V. Kurchatov and later M.A. Lavrentyev supported Budker and his work and indulged his eccentricities. A.M. liked to recall the reaction of I.V. Kurchatov to the comments of the leading experts about Budker's idea for colliding beams. Their comments were different, but all extremely negative. Igor Vassilievich laughed and declared, "Therefore, there is something nontrivial in this. We have to do it."

When the Novosibirsk scientific center was being developed, M.A. Lavrentyev actively and strongly supported "promising" groups. The Institute for Nuclear Physics, which quickly became a large, internationally known institute, was a favorite of his. However, in the last years of his life, A.M. felt a harsh change in the atmosphere. Many initiatives for the Institute were blocked, and the work dragged.

As the rector of the Novosibirsk University, I had regular good contacts with M.A. Lavrentyev and tried to change the situation. Unfortunately, my efforts helped only to remove individual misunderstandings. The negative attitude toward A.M. persisted, and new logs were constantly added as fuel to this smoldering fire. In this situation A.M. felt himself to be defenseless and occasionally became depressed.

Nevertheless, he had much firm support — from his beloved Institute and his collaborators and associates who rallied to him, from international recognition, and from his audacious projects. And life went on.

---

[10]Trans. note: Gosplan is the abbreviated name for the State Department of Planning.

Life is Creating!

B.V. Chirikov

The first time I met Andrei Mikhailovich was at the end of 1947. At that time I was starting my second year at the newly organized physico-technical faculty of Moscow State University, which was later reorganized as the Physico-Technical Institute. The classes began in a partially constructed, poorly illuminated, and even more poorly heated building near the Dolgoprudnaya train station in the suburbs of Moscow. Andrei Mikhailovich taught seminars on physics. He was not yet even 30 years old. Thin, slim, and good-looking, with rapid, impetuous motions and gleaming eyes, a beginning physicist who was unknown at the time, he immediately attracted the students' attention because of his extraordinary enthusiasm and almost child-like admiration for the world of physics, which he attempted to open up for us quite passionately. Among ourselves, we called him "Red"[1] — on account of the strong impression that he made on us and on account of the color of his noticeably thinning hair.

Andrei Mikhailovich told us about many interesting things, not necessarily with any particular relationship to the syllabus for the physics

[1] Trans. note: In Russian, the nickname "Red" is also used to refer to circus and carnival clowns, on account of their red wigs and likeable behavior.

103

course. I remember several seminars about parametric resonance. He spent a disproportionate amount of time on this subject, and not casually. At the time he was occupied with theory for the phasotron — an enormous (in those days) proton accelerator, constructed at the Laboratory at the Big Volga (which is now the Joint Institute for Nuclear Research in Dubna). One of the problems was the mysterious loss of the beam, which had been discovered on a similar accelerator by American scientists at Berkeley. Initially the beam would be successfully accelerated, spiraling within the narrow gap of the giant electromagnet, but suddenly, without reaching the edge of the magnetic, the beam would be entirely lost, precisely at the point where the logarithmic radial derivative of the magnetic field strength became equal to the characteristic value of $1/5$.

— "Why?" Budker asked us, spreading out the fingers on one of his hands. "Where does this magic number 5 come from? Is it the number of fingers on a hand?"

It seemed that what most amazed him was the value of the number itself, and he wanted to transmit to us this feeling of being amazed by the enigmas of nature, even by not very big enigmas.

— "The reason lies entirely in parametric resonance," he continued. Then he gave us an enthusiastic explanation about its detailed mechanics.

Perhaps it was from this time that I became keenly interested in oscillations and resonances, which determined to a significant degree my scientific destiny many years later. At that time, however, our paths diverged. Budker's seminars finished, and I underwent training and then worked for some time at Laboratory No. 3 (now the Institute of Theoretical and Experimental Physics in Moscow). Andrei Mikhailovich Budker worked at Laboratory No. 2, the famous "Number Two" (now the I.V. Kurchatov Institute of Atomic Energy).

Much later I learned that Andrei Mikhailovich Budker had earlier graduated from the physics department of Moscow State University in June of 1941 and had immediately gone into the army, and that after the Victory in Europe, his unit had been transferred to the Far East. Afterwards he told me how dumbfounded he had been by the brief information in the newspaper about the atomic bombing of Japan. The words "atomic bomb," which were so mysterious in those days, he really did understand. His reaction was instantaneous — at any cost he wanted to participate in the solution of the "atomic problem" in the U.S.S.R.

Thus, he went to Moscow and talked with A.B. Migdal, the head of the theoretical department of Laboratory No. 2, who would be his future teacher. Migdal asked him some questions about nuclear physics, a subject Budker had soundly forgotten during the war. One question was, "What is the spin of the deuteron?" Naturally, he did not remember, but he quickly thought up an answer: "Zero or one." Migdal liked his answers, and Budker's fate was decided.

My next encounter with Andrei Mikhailovich occurred casually in the spring of 1954 near the entrance to the main building of Laboratory No. 3. He remembered me from the Phys-Tech days, and we started to talk. Budker had come to the seminar with criticisms of a new type of accelerator that was being actively discussed then and that made everyone enthusiastic — viz., the strong focus accelerator. Budker's idea was that if the number of particle oscillations near the equilibrium orbit in a single turn is large, the primary phase of the oscillations must somehow be "forgotten" and, consequently, the amplitude of the oscillations will grow diffusively, and this, in turn, would lead to rapid loss of the particles. Those who were enthusiastic about the new accelerator idea disapproved of Budker's criticism, calling it "Budkerism" and thinking (not without reasons) that they had succeeded in overcoming his criticism and proving the prospects for the new accelerator.

I recall one more of Budker's arguments on this subject with Matvey Samsonovich Rabinovich,[2] one of the pioneers of strong focusing. It occurred later, in the large hallway of one of the buildings of Laboratory No. 2. Budker walked up and down the hallway, from one end to the other, proving that all the particles in such an accelerator will necessarily be lost. Rabinovich objected, saying that, to the contrary, it would be no problem to keep them near their equilibrium orbits as long as desired. Andrei Mikhailovich always liked vivid comparisons:

— "Listen," Budker said to me, "he wants to collect all the air molecules in one corner. But how can it be done, Musya?"

— "With the help of eigenfunctions," answered Rabinovich imperturbably. He tried to explain something, but Budker did not listen.

[2]Editors' note: M.S. Rabinovich (1919-1982) worked on the physics and technology of accelerators at that time.

Who, then, was correct? This question intrigued me and became one more motivation for my future interest in nonlinear dynamics. Within linear theory, Rabinovich was of course correct. Here, it was necessary only to avoid resonances, exactly the same topic that Budker had once studied in the phasotron. Now, however, there were too many resonances. Andrei Mikhailovich undoubtedly understood this, and the essence of his argument was whether linear theory was adequate for this problem.

As it turned out later, the dynamics of nonlinear oscillations is much more "cunning." Furthermore, at the same time that these energetic debates were taking place in Moscow, "numerical experiments" were being carried out at the European Center for Nuclear Research (CERN), on the outskirts of Geneva, where such an accelerator was also being developed. These computer experiments (which, incidentally, were some of the first numerical experiments in physics) were still unknown in the U.S.S.R. They showed that, even with very weakly nonlinear oscillations, diffusion occurs that is identical to what Budker had predicted, but whose mechanism he had been unable to explain at the time! But all this happened later. It now seems to me that, at the time, not only were his specific objections not very clear, but also Andrei Mikhailovich simply did not like this resolution of accelerator problems by means of strong focusing. I suppose that he considered it insufficiently revolutionary, and he had his own, much more grandiose, project for a new accelerator.

On that spring day in 1954 I also asked Budker about his other work. I was intrigued by his few phrases, which were unusually laconic for him, and I "demanded" explanations. We arranged to meet especially for this purpose.

Before telling me about his work, he said very seriously, "These things are strictly classified, and in talking about them, I am definitely violating instructions. Don't let me down." Naturally I promised that I would not do so, and I did not. (At the end of the 1950's, his work was all declassified, following the famous report by I.V. Kurchatov in England.) The content was about controlled thermonuclear reactions and about Budker's new suggestion (a trap with magnetic "plugs") that he was proposing in opposition to the scheme of Sakharov and Tamm, which appeared to have serious difficulties. He also described his accelerator based on a powerful relativistic electron beam, stabilized by its own radiation.

We talked for a long time. Andrei Mikhailovich naturally told me about everything, not merely to satisfy my curiosity or even my quite natural scientific interest — but because he needed new collaborators. And, I became infected with excitement.

Also, he asked me about my work. I was then involved in experiments with mesons on the phasotron, the same device whose theory Budker had been developing several years earlier. Out of all of my narrative, he was impressed only with my project (which was never to be realized) for reducing the multiple scattering of mesons in the air. I wanted to set up a long "intestine" of thin rubber (which was immediately renamed by Budker in another way) and fill it with hydrogen at normal pressure. Andrei Mikhailovich always valued inventiveness in people, correctly thinking that without it, nothing could be accomplished, not only in experiments, but also in theory. Anyway, I was suitable to him, and he enlisted me into his future experimental group — still unofficially, since he was merely a member of the theoretical department.

My transition to Budker was not painless. Many people tried to dissuade me from doing so.

— "It isn't physics at all, just boring technology," said one of my colleagues.

— "A real physicist should occupy himself with detectors, not with accelerators," said another. "Even supposing that Budker will build his accelerator in ten years, during this period the physics will already be done on usual accelerators."

— "Budker has a terrible personal character," warned a third.

And one physicist expressed his thoughts even more vividly: "Budker is a visionary. One should do a ruble's worth of work, rather a million rubles' worth, like this dreamer wants — also, however, not a kopeck's worth, but a ruble's worth!"[3]

In those days it was difficult for me to understood all this intricacy of opinions and assessments. I was simply interested in Andrei Mikhailovich's ideas and I trusted him, for which I was never sorry later.

---

[3]Trans. note: The meaning here is that one should be neither overly ambitious nor insufficiently ambitious.

Budker also paid a price.

— "Andrei, don't you know that you shouldn't lure away the house-maids of others?" an elder colleague said sternly to him.

Andrei Mikhailovich tried to justify himself, but did not change his decision. And here I was — in the enormous office of Igor Niko-layevich Golovin, deputy to I.V. Kurchatov (first deputy, apparently, judging from the size of this room) — the same size as Kurchatov had, both rooms having the same waiting room. Golovin actively supported Budker's research and helped him greatly with the organization of the experimental group. He received me very kindly and welcomed my decision to join Budker's group, but at the end of our conversation sternly asked me in his thunderous voice what, in fact, were the concrete specific thoughts that I had on how the wonderful ideas of A.M. Budker could be realized. I honestly admitted that I did not yet have any concrete thoughts and, besides, I was acquainted with these ideas themselves only in general, but that I hoped the thoughts would appear during the course of the work. Apparently Igor Nikolayevich was satisfied, and I was now officially enlisted.

Meanwhile things were not going too well for Budker. His idea of creating and heading an experimental group caused stormy debates and objections.

— "How can you appoint a theorist, who can't even drive in a nail, as the leader of an experimental group?" some people said.

This was an obvious exaggeration. Already before the war, Andrei Mikhailovich had trained at an institute that did research on the coloration of fur, where he had performed spectroscopic measurements. He had been quite impressed that, in a field seemingly so far removed from physics, there had been used (and very successfully!) a beautiful theory of human color vision and the most modern optical instruments.

Nevertheless, Budker was a theorist.

— "Besides," said one influential official, "he has no organizational abilities. For example, recently I asked him to help me choose a dacha.[4] What happened? It turned out that he was absolutely of no help."

— "Why the hell should I care about his dacha," gloomily remarked Andrei Mikhailovich, recounting this episode.

"The Beard" (Igor Vassilievich Kurchatov) hesitated, too. At last, there was found a "compromise": an experimental physicist from

---

[4]Trans. note: A summer cottage.

Leningrad was specially invited to be the director of the new Section No. 47, created for the experimental realization of Budker's ideas, while Andrei Mikhailovich was left as a powerless scientific leader.

He was extremely depressed. We, his unofficial staff, tried to calm him. The most important thing was that at last this section had been created for carrying out his ideas. Gradually the work got going, and the first interesting results appeared. Finally, it was admitted that it would be expedient for Budker to lead his own investigations by himself. Things became even better. The section grew rapidly and soon was reorganized as the Laboratory for New Acceleration Methods.

I came to Andrei Mikhailovich just at the time when he was finishing the elaboration of his thermonuclear proposal, and he involved me in this work. What a wonderful experience it was! He looked at a multitude of various options and considered all the different aspects of a problem. During the course of the work, what he needed most of all was a listener and a critic, rather than an assistant.

One day he came in, very excited, and joyfully exclaimed, "I have thought up the ideal trap!" It had a rather complicated structure, with different fields that varied in time and with plasma being pumped back and forth from one part to another. I was successful in "exposing" Andrei Mikhailovich.

— "What a pity," he said after thinking a bit. "A beautiful idea has failed. I had a feeling there was something wrong with it."

Another time he appeared, very upset, munched on his lips a little, and vaguely drawled, "Yessss...." He then scrawled a few formulas all over the first piece of paper he found.

— "See, here comes a logarithm." He explained that the loss of particles from the trap turned out to be much larger than he had initially expected. He calmed down a bit, then suddenly concluded:

— "This is the way to work! At first, approximately, by orders of magnitude. Then we'll be specific and carefully check everything. However, the effect of multiple scattering is clear. We have to think what to do about it."

Andrei Mikhailovich immediately noticed an important advantage of his future thermonuclear reactor — the possibility of directly transforming nuclear energy to electrical power, without intermediate thermal and mechanical stages. He investigated the details of this interesting question. He understood that all of this was to some extent premature, and so he liked to tell the anecdote about an inventor who developed a regulator for the *perpetuum mobile* so that it would not become too powerful.

Budker also paid attention to another problem. Particles are confined in the trap due to conservation of an adiabatic invariant. However, this conservation law was satisfied only approximately. Was this sufficient to confine particles during many millions of oscillations, until thermonuclear reactions would be attained? No one at that time could answer this question. But Andrei Mikhailovich found his own solution.

He clearly understood (and explained to us) that, from the theoretical point of view, the problem was a very subtle one — nothing could be done simply by evaluating the order of magnitude. Hence it was necessary to carry out a model experiment, simple and convincing, an *experimentum crucis,* which was always his ideal and dream. And he thought one up! Its sweet point was that Andrei Mikhailovich succeeded in avoiding the problem of the injection and capture of charged particles in the trap, which was technically difficult at the time. He suggested filling the trap with tritium and using the electrons that are produced when it decays. He also thought out a very easy and reliable technique for measuring the confinement time for the electrons in the trap. The main thing is that the experiment was elegant, and it was immediately carried out by Stanislav Nikolayevich Rodionov, Budker's student and one of his first colleagues. The work gave a convincing proof for the possibility of confining the particles for a long time in Budker's trap, as we now call it.

Andrei Mikhailovich was very proud of this experiment, no less, it seems to me, than of his thermonuclear ideas. He continued to tell everyone about it for a long time, recounting from the very beginning the detailed circumstances and the entire sequence of his thoughts. He became especially excited when anyone thought that the results of the experiment were obvious or said that everything was clear without it. He tried to prove that this experiment was of great importance not only for the realization of controlled thermonuclear reactions, but also for the fundamental issue of adiabatic invariants.

In general, Budker's ideal, as he told us many times, was the experiment of Pasteur, which simultaneously solved a fundamental problem of biology (the impossibility for life to be self-originating) and which had great practical significance for people (the pasteurization of food).

I myself at that time recalled the argument between Budker and Rabinovich about the mysterious diffusion of particles in an accelerator. In a magnetic trap, particles also perform oscillations, just as in an accelerator, but in the case of the trap, linear theory is not at all appropriate. However, what happens if...? Rather quickly I was successfully able to do an initial calculation, the results of which to some extent were even in agreement with the experimental results of Rodionov. Following my first report at a seminar, Andrei Mikhailovich remarked, "You must have been very brave to dare to compare such a theory with an experiment." I went on to become immersed in more detailed calculations and, later, also in numerical experiments. Budker's problem, as it is now called, turned out to be extremely fundamental and interesting. Subsequently, many researchers, including mathematicians, worked on this problem, although it is still not completely solved. Meanwhile the thermonuclear idea of Andrei Mikhailovich Budker began to take on a life of its own, and it currently constitutes one of the principal lines of research and investigation in this area.

Much later, when he was already the director of the Institute of Nuclear Physics in Novosibirsk and headed up a large scientific group as well as his own school of physicists, Budker frequently recalled his time in the army. He thought that military service had greatly helped him understand interpersonal relationships and also the dynamics of managing collective creativity, which he studied and put into practice with the same energy and interest as he studied the dynamics of elementary particles. Budker remembered his commanders with gratitude, especially his first one: "He was exacting, strict, and absolutely fair!" Many times Budker quoted this line from the military regulations: "A commander must make a decision."

— "Nothing is said in the regulations about making the best decision or the correct one," he emphasized. "It is the absence of any decision, passivity, and confusion in critical situations, that is much worse."

Andrei Mikhailovich always followed this rule.

One such critical situation occurred in 1956. The Laboratory for New Methods of Acceleration at the Institute of Atomic Energy, which was already a rather large group headed up by Budker, was working enthusiastically to realize Andrei Mikhailovich's primary idea — the stabilized electron beam — which in his mind would open up a totally new vista for accelerator technology. Impressive success had been obtained in creating high-power electron beams. However, it was becoming more and more evident that the problem as it had been formulated was much more complicated than it had seemed at first, so that the ultimate goal of a stabilized electron beam appeared to be receding farther away, rather than drawing near.

What must be done? Of course, we could be carried along by inertia, turning into just an ordinary laboratory that obtains not-so-bad second-rate results about the physics and technology of beams, which have different, perhaps significant, by-applications.

But this was not what Andrei Mikhailovich had dreamed about in the torment of creating his group! He could only be satisfied by a cardinal, revolutionary decision, and he found it — colliding beams.

It was not very easy for Andrei Mikhailovich to make this decision, because it was not his own idea. It had already been discussed at scientific conferences, and there was even a project for such a device. However, Budker clearly understood that, on the one hand, colliding beams would indeed revolutionize accelerator technology and high-energy physics, whereas on the other hand, the only place in the Soviet Union that could carry out this project was his laboratory, with its foundation of extensive experience in working with high-power electron beams, which would then not go to waste. And so he made the decision. It defined the fate of the Laboratory, and also that of the Institute of Nuclear Physics in Novosibirsk, which was set up on the basis of this laboratory soon thereafter. I can only add that this is still the main subject at the Institute of Nuclear Physics and that our institute is the only place in the U.S.S.R. where such devices are in operation.

Academician I.V. Kurchatov actively supported Budker's decision and rendered great assistance in carrying it out. At first it was assumed that this would involve a considerable enlargement of Budker's laboratory in order to provide the scope and pace necessary for the development of the work. However, already after one year, the general atmosphere changed dramatically, and a new critical situation arose — the Laboratory was in real danger of becoming a bird with its wings clipped.

Once again, it would have been possible for us to submit and to continue "tilling" a small but comfortable "garden" in Moscow, "planting" beautiful "flowers," "laying out" and "rearranginging" our "flower beds and mounds." But A.M. Budker was not this kind of person! So without hesitation he accepted the offer (a very timely one!) of M.A. Lavrentyev and I.V. Kurchatov to organize an Institute of Nuclear Physics in Novosibirsk, at the recently established Siberian Division of the U.S.S.R. Academy of Sciences. Above all, Siberia attracted Budker with its grand scale, its grand scale in everything. Indeed, the spaciousness of Siberia was what he needed to be able to give expression to his vigorous energy and initiatives. It was not without reason that academician I.V. Kurchatov, in a speech at the general meeting of the U.S.S.R. Academy of Sciences in which he supported the organization of the Siberian Division, made the particular statement, "Our institute is sending to Siberia its most active group — Budker's laboratory."

Several years later, after the first building of the INP was completed, Andrei Mikhailovich moved once and for all to Akademgorodok[5] in the suburbs of Novosibirsk. He loved this region and remained its ardent champion, "meddling" in all aspects of the life of this city of science whenever he had opportunity.

Budker also actively participated in organizing Novosibirsk State University, modeled after the famous "Phys-Tech" — Moscow Physico-Technical Institute, where he had begun his pedagogical activity more than ten years earlier.

At the end of September, 1959, I went with Andrei Mikhailovich to Novosibirsk, at that time only on a mission, to attend the opening of Novosibirsk State University. There was no solemn ceremony (that took place much later), just the first day of classes, the first ever lectures at NSU, held in the assembly hall of the school building (now School No. 25), that was donated to NSU until its first building was constructed. The "Phys-Tech" had begun in almost the same way!

Two days before the opening, we left the hotel in the evening for a stroll in downtown Novosibirsk. Samson Semyonovich Kutateladze and someone else, apparently one of his colleagues, joined us. We walked along Krasny Prospect to the river, came to Kommunalny Bridge, crossed to the other side, and went down to the edge of the water. It was already rather dark and cold. In front of us flowed the lead-colored water of the powerful Siberian river Ob'. Suddenly, at this moment, Andrei Mikhailovich decided that he just had to dive into the Siberian river! We tried vainly to dissuade him. He quickly got into the water,

[5]Trans. note: "Academic City."

swam a little, and then rushed back to the shore very happy, swinging his arms energetically and jumping up and down trying to warm himself. This was really his baptism in Siberia!

Andrei Mikhailovich did not fall ill, but... he completely lost his voice. We had to treat him immediately with hot milk with soda. Fortunately, everything was all right, and at the scheduled time he gave a lecture about physics — the second lecture ever given at NSU. The first lecture, concerning mathematics, was presented by academician Sergey L'vovich Sobolev. The new university was now launched into orbit.

The following day, Budker returned to Moscow, while I remained behind to continue the course of lectures. I had to move urgently to Novosibirsk. It was my first teaching experience — difficult, but very interesting.

Budker more or less regularly came to Novosibirsk for a few days at a time. He had many things to do here — for the Institute was being built. But in addition, he would give one or two lectures, and he would go through with me the details of the material to be presented next, which I would later have to explain to the students on my own.

This was a wonderful experience for me! To "go through" almost all of physics one more time with a teacher like A.M. Budker! He could always find an unusual point of view and a novel way to consider seemingly well-understood issues. He placed great significance on the proper interpretation of physical laws — on the philosophy of physics, I might even say. A positivistic phenomenology with its conventional agreements was absolutely alien to him. Above all, he searched for the mechanisms of real physical phenomena, and he did it brilliantly!

It was at this time that Andrei Mikhailovich had the idea to begin teaching physics in the first year directly from the theory of relativity. He thought that first teaching classical concepts of physics to students and then promptly reeducating them the next year was wrong.

— "Future physicists must be directly trained in relativistic thinking," he said.

There was another side to this issue. Modern physics is quite unusual from the point of view of our ordinary notions, even the notions of physicists themselves. Academician Lev Davidovich Landau expressed this very well at one of his meetings with the Phys-Tech students, when he stated, "The power of human reason is also exhibited in the fact that we are successfully able to study phenomena that are already beyond our ability to imagine."

But only study was not enough for Budker. He was not only a researcher, but also an inventor in the highest sense of this word, an inventor on the frontier of modern physics. Landau had good reason when he once called him a "relativity engineer." Probably there was some irony here, but Budker was not offended. On the contrary, he was very proud of this "title" and frequently quoted Landau's words.

However, an inventor cannot work without images. As he mulls over in his mind the hundreds of options, while trying to circumvent or "outwit" the inviolable laws of physics, which Andrei Mikhailovich respected as sacred, the inventor must not only have an absolutely clear understanding, but must also "see" the bizarre and invisible "phantasmagoria" of modern physics. Budker proved that this process was quite possible for him to do, and that it could also be taught to others.

A few years later, at a meeting at the Ministry of U.S.S.R. Higher Education, I had to describe our experience in teaching the theory of relativity early. The audience received my report with suspicion, not to say hostility. One of the strongest opponents of such innovation even jumped up from his seat and shouted, "Who gave permission for such experiments on students?" But after a few more years, Budker's idea gained widespread acceptance and was even recommended by the Ministry for physics majors at all universities in our country.

Yes, Andrei Mikhailovich fell in love with Siberia. He entertained no thoughts about returning to the capital.

— "Here is my home, here I'll die, here I'll be buried," he used to say with a smile.

Here also is his Institute — his most important and beloved creation, to which he dedicated his life. This beautiful Institute of Nuclear Physics continues to grow as a living memorial to this great Soviet physicist, whose wonderful motto was: "Life is Creating."[6]

[6] Author's note: This was Budker's antithesis to a very popular motto of those days, "Life is Struggle."

The Unforgettable Years

B.G. Yerozolimsky

It has been well-known for a long time that anyone who writes reminiscenses about certain destinies in life or about personalities will inevitably, in one way or another, necessarily write about himself. The author of this reminiscence asks in advance to be excused for being unable to avoid the same thing, because all of his strongest recollections about A.M. Budker are inevitably connected with the fact that his own life became interwoven, as fate had it, with the life of this remarkable man.

I met Andrei for the first time at the end of the 1940's at the famous Laboratory No. 2 of the Academy of Sciences, established under the leadership of I.V. Kurchatov, whose name was given to it when it became the Institute of Atomic Energy.

There was a special atmosphere at the Kurchatov Institute in those years: people worked enthusiastically, staying in the laboratory for 24 hours at a time, with enthrallment and pleasure, applying all their efforts in order to solve, as quickly and as well as possible, scientific and technical problems that I.V. Kurchatov posed for the group.

Budker's personality, according to my memories of those wonderful years, fit in perfectly with this scenario of shared intense work and enthusiasm. I remember well how he paced back and forth along the corridors of the main building of the Laboratory (where in those days the theorists were also housed) — a young man, but already bald, either completely immersed in his thoughts or excitedly explaining to someone the latest idea that had dawned upon him.

Already by that time Budker clearly exhibited two particular features of his creative "handwriting." First, though at that time a "pure" theorist, he disliked working at a desk or with books — he had a magnificent memory and was quick on the uptake, and despite his dislike of carefully working through books, he already possessed by then wide-ranging and deep erudition, which, combined with an enormous creative imagination, gave him the capability to generate ideas that he later "polished" in conversations with everyone he ran into, in the theorists' offices as well as in the hallways of the Laboratory. This "method" of work was to become his main "technology" for creativity during his later years when he had become the leader of a large group of scientists.

Secondly, even at the very beginning of his creative path, as he worked among the "pure" theorists who investigated fundamental problems of nuclear theory under the leadership of A.B. Migdal, he was clearly attracted to problems with technological aspects, which included coming up with inventions, which, however, were based on a deep understanding of physical laws in what were sometimes the most abstract and fundamental areas of knowledge, such as electrodynamics and the theory of relativity. Subsequently this aspect of Andrei's creative nature as a scientist would become the determinative one for him and would also essentially typify the entire Institute that he established.

In these long-ago years, no one who closely knew Budker could suspect the true scale of his personality. They regarded with some condescension the cascade of ideas that he pressed upon his colleagues, as the fantasies of a somewhat eccentric young man. Some even had an hostile attitude, especially later when Budker, as he acquired authority and became acknowledged, began to lay claim to leadership over a group of experimentalists for implementing his ideas and projects. Here I must particularly emphasize that only a very few people were able to discern in him the ability to lead people, especially experimentalists, even on the scale of a small laboratory, to say nothing of an institute. It is strange how firmly and unchangingly people maintain their stereotypes about opinions and ideas they once had. Even many years later, when Budker had already become the director of the Institute of Nuclear Physics in Siberia, I frequently heard his former colleagues make sceptical references about his ability to be a leader.

At that time, the beginning of the 1950's, the opinion about Budker was firmly established: he was considered to be a cheerful and witty person, a marvelous volleyball player (in those days he was an essential member of the Institute's volleyball team), an excellent dancer (he even earned money while a student by teaching dancing), epicurean in his attitude toward life and its good things, undeniably gifted, but with noticeable traces of provincial upbringing. People always felt ready to enjoy listening to him tell funny stories (which he told in a most expressive and humorous manner) or to argue with him about some scientific, philosophical, sociological, or political subject; but to give him the responsibility for practical activities at a laboratory, how could this ever be possible! Even I.V. Kurchatov, so it seems to me, was under the influence of this opinion about Budker for a rather long time, not daring to appoint him as the leader of the newly established section that was specially set up for the experimental development of one of Budker's ideas. A few years later, Kurchatov, who was himself a remarkable scientific organizer, appreciated Budker's full measure and rendered great assistance to him when a special large department, the Laboratory of New Acceleration Methods, was set up at the Institute of Atomic Energy, again when the program to create facilities for electron-positron colliding beams was being carried out, and finally when the Institute of Nuclear Physics was being organized.

My direct work with Budker concerns precisely this period (1955-1962).

It all started one evening when Andrei dropped by to see P.E. Spivak,[1] with whom I, along with others, was then working, and told us about an elegant and what we thought seemed like an extremely simple and efficient idea for accelerating plasma blobs. The main point was to inject a burst of gas quickly between two metal bus-bars to which is

---

[1] Editors' note: P.E. Spivak, a corresponding member of the U.S.S.R. Academy of Sciences, was at that time head of a laboratory at the Institute of Atomic Energy.

applied a high voltage in vacuum and which are situated in a strong transverse magnetic field. A current generated after breakdown of the gas will move under the influence of the magnetic field, and thus the "bridge" of plasma will accelerate along the bus-bars, like a train along railroad tracks. Simple estimates, made the same evening, showed that under experimental conditions that would be easy to attain, with a length of about half a meter, one could expect to accelerate a blob of hydrogen plasma with the protons attaining an energy of about 1 MeV.

A small table-top proton accelerator — an experimentalist's dream! Inspired and excited by this idea, we decided to start working to implement it. From this evening on, our collaborative intensive efforts started running for days and nights and weeks and months.

Nothing came of this project. After fooling around with it for more than a year, we had to give it up. Nowadays this is generally known as a trivial thing, but at that time we had no idea of the insurmountable difficulties that awaited us along the way and that were related to instability of the plasma: it actually slid some tens of centimeters along the rails, but then went to pieces before attaining the desired energies.

Budker was inexhaustible in inventing things, suggesting new hypotheses one after another that could explain our failures and then inventing the corresponding modifications to be made in our experimental device. Budker, who had the highest esteem for the wealth of experience and very keen experimental intuition of P.E. Spivak, followed our efforts quite attentively and frequently took a hand even in the practical work. All in all, was this not the first experiment to be done under the guidance of his ideas? It is interesting to note that this, apparently, was one of the first Soviet investigations of the instability of plasma. Also, this was a very simple experiment, out of which began the long road that Budker and his colleagues traveled along in their studies of plasma physics.

Around the same time there was conceived the famous idea of magnetic mirrors ("stoppers"), which came to Budker as a result, I think, of discussions with P.E. Spivak about his project for making a device to measure the charge-to-mass ratio for positrons. Spivak consulted with Budker about methods to calculate the orbits of positrons in a longitudinal magnetic field that increases in strength along its direction, which P.E. Spivak wanted to apply in order to "collect" particles from a large volume to bring them to a detector window. At the time Andrei was thinking about ways to achieve magnetic insulation of a hot plasma, which is necessary for solving the problem of controlled thermonuclear fusion. Suddenly he had the idea of using a magnetic field configuration of this type to create a magnetic barrier for the plasma; devices of this sort are nowadays called open magnetic traps.

In the spring of 1956, at the invitation of Budker and with the blessing of I.V. Kurchatov, I moved to the newly organized section whose purpose was to figure out a way to implement Budker's idea of the so-called stabilized electron beam. In those days this idea seemed to be the only real means for being able to accelerate protons up to energies of about 100 GeV. It was just at this time, following a long period of arguments, conflicts, and events of an occasionally dramatic nature, that Budker had been confirmed in his appointment as head of the section, with all the work proceeding entirely under his guidance. The directorates of the agency for atomic energy and the agency for state security held out furiously against this appointment — a Jew and not a party member, as the head of such an important section! Such a situation was intolerable for them. And there had even been an interval of time during which Budker had been deprived of a security clearance, and only the steadfast support and defense from I.V. Kurchatov had prevented Budker from being altogether dismissed from the Institute. Yet finally his appointment was confirmed.

This was an unforgettable time of youthful excitement and enthusiasm, of very heated arguments, of ideas big and small that seemed to pour out of a cornucopia, of selfless work by everyone (without exception), all of us having been gathered together by Budker. Though young and cocky, he was acknowledged by us all as the undisputed leader of the new direction in accelerator physics.

Several years were spent working on these projects. The primary goals were not fully achieved, because along the way there arose certain particular obstacles, also related to the same instability of plasma carrying a high current, which were difficult to overcome. Nevertheless, in the course of these investigations, important results were obtained that significantly advanced the technical capabilities for creating and controlling high-current electron beams at high energies, as well as enhancing the understanding of the physical processes in relativistic electron plasmas. We developed special high-current pulsed sources of electrons, as well as fast systems for beam control, with innovative generators in which the voltage could change by hundreds of kilovolts in nanoseconds. In the devices built during these years, electron beams with a current of several tens of amperes, circulating in vacuum, were obtained — for the first time in the whole world.

Naturally, due to the success we had during the course of these investigations, it became possible to carry out a whole range of projects for unique devices. For example, we were the first in the Soviet Union to build an ironless pulsed synchrotron, with an electron energy of 200 MeV.

However, perhaps the main result of these creative efforts was the creation of a first-rate scientific group that was later able to shoulder problems that were even more serious in terms of their essential physics and enormous in terms of their scale. Without a doubt, most of the credit for this belongs to A.M. Budker. This man who could tell all kinds of stories and anecdotes with such exuberance and animation and who could construct hypotheses about why the birth rate for boys increased after an epidemic or a war and who could theorize about the features of space near "black holes," this man turned out to possess an outstanding talent for organizing scientific research.

During those years, we made many mistakes and oversights (and learned from them!). Budker himself could be biased and unfair or at times childishly naive. For example, while giving a presentation at a large assembly, Budker was once driven to exasperation by a venerable academician who wearied him to death by asking typically hair-splitting and meticulous questions. No longer able to contain himself, Budker said, "Listen, it's well-known that a hundred wise men would not be able to answer all the questions of one...."[2] Here he stopped short, having remembered how this famous proverb ended, which would have been quite inappropriate to say publicly. By that time everyone present was already laughing loudly.

Yet, along with these failings he had an extraordinary gift to attract people, enthrall them, and convince them. In addition, he had a special intuition for discerning talented people, for whom he was always searching among the youth of the Institute and the students, to involve them in his work.

---

[2] Trans. note: The final word of this proverb is 'fool.'

It was precisely at this time when Budker assembled a brilliant "Pleiades" of physicists and engineers, such as A.A. Naumov, V.S. Panasyuk, B.V. Chirikov, E.A. Abramyan, L.I. Yudin, V.I. Volosov, A. M. Stefanovsky, I.M. Samoilov, and S.N. Rodionov, as well as some rather young persons — L.N. Bondarenko, Yu.A. Mostovoy, A.Kh. Kadymov, and many others. With these people Budker achieved his first successes; with them, first a large department at the Institute of Atomic Energy (called the Laboratory of New Acceleration Methods) and later the Institute of Nuclear Physics of the newly organized Siberian Division of the U.S.S.R. Academy of Sciences were born.

Budker thought long and hard and talked a great deal about what the principles for organizing a scientific laboratory should be like in order for the scientists to work most efficiently and for establishing and nurturing a creative atmosphere and morale within the group.

I should emphasize the following three organizational principles from among the many that he understood:

1. The priority of scientists at the Laboratory (i.e., the Institute). No matter who — scientist or engineer —, creative people are the foundation of the scientific group, and all the technical, service, and administrative staff members (including those at the management level) must perceive that the goal of their activities is to help the scientists. Even though the truth of this principle is quite obvious, unfortunately it is truly implemented much too seldom at our scientific research institutions.

2. The necessity (or, at the least, the urgent desirability) of complete unanimity among the leading staff (e.g., the Scientific Council or any other consultative body) in making important creative or organizational decisions. This principle is quite nontrivial, and carrying it out is rather difficult — everyone is different and, as a rule, has different points of view about all issues. So, Budker spent hours, even days, in heated discussions, trying hard to persuade people and bring them to a concensus opinion. Following a heated argument, if he could not make someone (whose opinion he took into consideration, of course) change his mind, he would be extremely depressed and would leave upset, his head hanging down. He thought that making decisions by force or by majority vote was a sure way to destroy the unity of the group.

3. The refusal to compromise when resolving organizational problems (especially moral and ethical ones), regardless of their magnitude and significance. "You have to understand," he entreated his colleagues and his students, "in these issues, nothing is trifling." Many years later I happened to attend a meeting of the Scientific Council at the Institute of Nuclear Physics and was once again surprised at the enthusiastic interest with which they discussed distributing bonuses to members of the accounting staff (individually!): the tone and the level of involvement were set by Budker, who debated each candidate and insisted on the principle of unanimous decision.

In 1958, when the decision was made to create the Institute of Nuclear Physics in the Siberian Division and A.M. Budker was appointed as its director, a qualitatively new era began in the life of the group he led. Constructing the Institute; organizing the administrative and managerial staff; admitting new employees (mostly these were young people, about to graduate from Moscow State University and Moscow Physico-Technical Institute); hiring support staff, foremen, laborers, and engineers; acquiring machinery and equipment — it is hardly possible to list everything.

And it was then that the completely unexpected organizational skill of Budker fully manifested itself. Whatever he became involved in, he did so enthusiastically — from the architectural appearance of the future Institute to the brand names for equipment to be acquired, from the hiring of technicians to the search for a candidate for the position of deputy director. For the latter, he managed to find and persuade a remarkable engineer and leader, A.A. Nezhevenko, who was the director of the Novosibirsk Turbogenerator Factory and also a member of the regional committee of the Communist Party of the Soviet Union, to switch over to the Institute. Nezhevenko was the one who oversaw the construction of the Institute. To his final days, he was one of its main administrators and leaders.

I recall very well how Andrei, with his typical enthusiasm, explained to us that at the Institute there would be an obligatory large table for the meetings of the Council of senior members, and the shape of the table would necessarily be *round* because there must be no hierarchy or inequality among the participants. He put it like this: "The director must only be a senior among equals." Everyone who has ever been at the Institute of Nuclear Physics knows this huge Round Table very well. Here, in a free and democratic atmosphere, over a cup of coffee, are discussed the current problems of the Institute — from small details of everyday life, to plans for a new accelerator.

Practically at the same time as the new Institute was being organized, Budker began to develop in his department a line of research that was destined to become one of the main thrusts of the Institute of Nuclear Physics, one that would bring world fame to its creator as well as to the Institute. I refer to colliding electron-electron and electron-positron beams.

The brilliant scientific and technical achievements that Budker and his colleagues obtained in this field are well-known, and so I will not dwell upon this great and glorious chapter of the history of his scientific activity. I will merely mention that the idea of colliding beams with their large energy gain in the center-of-mass frame of the colliding electrons, in contrast to the case when the accelerated electrons hit a fixed target, occurred to Budker independently of O'Neill. Budker discussed it with us several months before the appearance of the preprint of the American proposal.

The devices with large accelerated currents of electrons that were already in operation at our laboratory at the time could serve as the basis for setting up a colliding beam system, and we thought that experiments on electron-electron scattering to check quantum electrodynamics at small distances could be carried out in something like two years. Therefore, in the winter of 1958, it was decided to undertake this work.

In fact, however, it took seven years to be able to carry out the first experiments. Still, the experience acquired during this work, along with a number of brilliant ideas by Budker, allowed the INP group during the next ten years to build the unique devices that vaulted the Institute into the ranks of the major world centers for high-energy particle physics research.

The way events transpired, after 1962 I no longer had the opportunity to participate in the work of the Institute nor, practically, to meet any more with Andrei.

I saw him for the last time in May of 1977. He invited me to the Institute of Nuclear Physics to be an official reviewer at a thesis defense. "And by the way," he said, "you'll see how we live here." The Institute profoundly impressed me with the scale and the elegance of its physics devices, as well as with its wonderfully creative and businesslike atmosphere that reminded me of the best years at the Institute of Atomic Energy in Budker's young group when Kurchatov was still alive. I saw that Budker had been able to implement in his Institute the scientific organizational principles about which he had thought so much and so deeply in the 1950's.

But Andrei was already seriously ill. He had had two heart attacks and walked slowly with a cane. However, when the Senior Council met at the Round Table, his voice sounded as youthful and enthusiastic as always during discussions with his students and colleagues who, despite his high titles and advanced age, argued and disagreed with him, also as always. Then he rapped on the table and exclaimed: "Silence! Discipline your thoughts! Why can't you understand?" And in this bearded old man with such sad-looking eyes, I saw the former young irrepressible Budker. Later that evening, when seeing me off from his home, he complained about the difficult conditions under which he had to work due to sharp disagreements with the leaders of the Siberian Division of the Academy of Sciences, and he began to elaborate on his dreams for setting up a new institute somewhere in the south of Ukraine. "Everything must be started over from the very beginning. I'll take along with me some good scientists and leave the others here to continue working without me.... If I were to suggest it to you and Piotr Spivak, would you go?" It turned out that he had already made detailed arrangements to establish a new institute somewhere near Krasnodar. But two months later he died.

Many years have already passed since that time, but the real measure of A.M. Budker, the person and the scientist, becomes more and more clear with each year.

At the Institute of Nuclear Physics, above the Round Table, which stands in the middle of the large conference room of the Scientific Council, there hangs on the wall a portrait of Budker carved out of wood. It is a marvelous portrait. It shows Budker as he was during the last years of his life — inspiringly talented and passionate, but at the same time grown wise in life and looking at us pensively.

Brushstrokes to the Portrait of Andrei Mikhailovich Budker

Ya.B. Zel'dovich

A.M. Budker will forever remain in my memory as one of the most outstanding persons in the group that formed in connection with the problem of nuclear energy and that then actively began to participate in research on elementary particles. By now I have already forgotten when and from whom I heard about him for the first time. I only remember admiration for the original thinking and the inventiveness of this young theorist.

The freshness of Andrei Mikhailovich's creative approach was amazing. Many in the older generation still remember the arguments over the "new physics" — relativity theory and quantum mechanics. Are these theories correct? How are they to be interpreted? How are they to be reconciled with classical physics? Budker had no such questions! For him, relativity theory was a ready and perfect instrument, a correct guide on the way to novel technology and novel devices. He accepted having different time clocks and different particle and charge densities in different coordinate systems as necessary and quite natural.

The breadth of Andrei Mikhailovich's creative interests should also be emphasized. Among his credits was the magnetic mirror ("stop-pertron") project — the original device for achieving thermonuclear fusion. Soon afterwards, he suggested the idea of using colliding charged-particle beams for investigating physical processes at record high energies.

To illustrate A.M. Budker's scrupulous nature, let me recall one instance: In his report at the general session of the U.S.S.R. Academy of Sciences, he specifically mentioned my comment that colliding beams have a great advantage in terms of energy, but at the same time he remarked that I considered colliding beams to be unfeasible in practice due to the difficulty of focusing them. That Andrei Mikhailovich picked up on the positive part of my statement was typical of his scientific bravery. At the same time, the initial (but also naive) pessimistic evaluations did not frighten him: he found ways to overcome the difficulties. Much later, in Novosibirsk, Andrei Mikhailovich raised the method of focusing to a higher level, by applying the principle of beam cooling, thus overcoming Liouville's theorem, which had seemed so firm.

However, I wish to write about the person, not about technical matters. Budker, who abruptly appeared on the scientific scene, "polarized" the public opinion of him held by physicists. Some spoke about an outstanding and talented man, whereas others called him a schemer and even an impudent provincial fellow, saying that "it is easy to write down formulas; you should instead try to do these things."

Here is another reminiscence that made an impression on me: We were talking about Budker with two highly respected academicians from LIPAN (the Laboratory of Measurement Instruments of the U.S.S.R. Academy of Sciences — the former name of the I.V. Kurchatov Institute of Atomic Energy). The academicians held opposite points of view. For ten minutes I literally had to hold one of them, the weaker one, by the arms, to prevent their argument from turning into a fist fight. However, none of the verbal arguments were decisive. What was decisive was the creation of the Institute of Nuclear Physics in Novosibirsk, where Budker's talent developed to its full extent. It was there that experiments with colliding beams were carried out. It was there that small radiation sources, needed for national economic purposes, were built and are now being mass produced. It was there that a highly skilled theory group, led by S.T. Belyaev, was built up.

It was precisely here, in Novosibirsk, where I came to a closer association with Andrei Mikhailovich, both personal as well as scientific. He told me about his difficult youth, about his interest in physics, and about the prizes he had received for best performance of ballroom dancing. He also talked to me about his life in Novosibirsk, about his yacht, and about fresh new ideas.

Just a few more words remain to be said, but these few words cannot be left out. I am unable to fully express my indignation concerning the fact that, at the general session of the U.S.S.R. Academy of Sciences, the motion was made not to re-appoint Andrei Mikhailovich as the director of the Institute of Nuclear Physics, which he himself had created! I learned that prior to this, an academic commission had been sent to Novosibirsk in order to examine the Institute of Nuclear Physics. To the credit of this commission, headed by academician Bruno Maksimovich Pontecorvo, it may be said that they supported Budker! The commission immediately became the target of attacks and was accused of unscrupulousness and favoritism, all this despite the fact that it would be hard to imagine how Pontecorvo, an Italian, could be like a godfather to Andrei Mikhailovich Budker, who had been born in a Ukrainian village.

Far be it from me to want to settle a score with somebody. I just want to understand why a quite healthy Budker had his first heart attack when he was 50 years old and died in the prime of his talent and intellectual powers when he was 59. There is no doubt that these troubles, described earlier, shortened Andrei Mikhailovich's life.

Our reminiscences would not be worth a farthing if we were to write them dishonestly, worrying about properly "censoring" ourselves, avoiding sharp corners and unpleasant episodes. This would not be what we were taught by Andrei Mikhailovich, who was sparkling, straightforward, and candid.

My Encounters with A.M. Budker

L.B. Okun'

"You know, I've found a mistake in *The Classical Theory of Fields* by Landau and Lifshitz," Andrei Mikhailovich said in the third minute of our first encounter. This conversation occurred in the beginning of the summer of 1953 at the VINITI[1] editorial office for physics. The editorial office had just recently been established, as had the VINITI itself, incidentally. While the first issue of the bulletin of abstracts was being prepared, Andrei Mikhailovich came to find out if he could be occupied writing abstracts: he had been dismissed from his job at LIPAN (Laboratory of Measurement Instruments of the U.S.S.R. Academy of Sciences) and needed money. He took from me about 30 articles on almost all areas of physics and promised to turn in abstracts by two or three weeks at most....

[1] Trans. note: VINITI is the Russian acronym for the All-Union Institute for Scientific and Technical Information.

The next time we met was at the end of the 1950's: I.Ya. Pomeranchuk[2] went to LIPAN to see the first of Budker's accelerators and took me along. Isaac Yakovlevich did not particularly enjoy visiting experimental devices. During the 12 years I had worked under his leadership, he had never gone to see either the reactor or the accelerators of the Institute of Theoretical and Experimental Physics. When we went to Dubna, he spent the entire time in discussions with physicists, indulging my enthusiasm for looking at "iron stuff." Therefore, the very fact that "Chuk" would make a trip to Budker's "accelerator show" was unusual and extraordinary.

Several accelerators in different degrees of readiness stood on the tables. Their reddish copper shone like polished mirrors. It was hard to believe that these copper things, about the size of a big pan for cooking preserves, could actually accelerate electrons. Andrei Mikhailovich spoke enthusiastically about the outlook for his forthcoming move to Novosibirsk.

In the autumn of 1967, I came on a business trip to Academgorodok. The Institute, created by Andrei Mikhailovich, made a great impression on me. It was quite different from anything I had seen before. There were talented physicists in a high concentration, youthful spirits, democracy, and no apathy. At the center of all the work was Andrei Mikhailovich — an inexhaustible generator of new ideas and projects. (In the *CERN Courier*[3] someone once wrote that Budker's ideas were like showers of sparks flying from a grinding wheel.)

Andrei Mikhailovich showed me a new tunnel and talked about the project for setting up colliding proton-antiproton beams with 25 GeV. (As C. Rubbia and D. Cline[4] later emphasized many times, this uncompleted project played a very important role in resolving the question of whether to carry out a CERN project that led to the discovery of intermediate bosons.) Over dinner in the old dining room of the House of Scientists, he began to urge me to move to Siberia: "You'll be able to travel abroad every year. You'll receive a 300 ruble salary increase."

---

[2]Editors' note: I.Ya. Pomeranchuk (1913-1966) was a theoretical physicist, an academician, the founder of a school in theoretical physics, and the recipient of several U.S.S.R. State Prizes.

[3]Editors' note: The *CERN Courier* is a magazine published by the European Center for Nuclear Research (CERN).

[4]Editors' note: Carlo Rubbia is a Nobel laureate in physics, and David Cline a well-known American physicist.

"Andrei Mikhailovich, I don't need money," I said proudly, but not very honestly.

"Well, then," he smiled. "You will give it to me."

In the evening, in his cottage, with a fire burning brightly in the fireplace, he showed me albums with photographs of visits by world celebrities to the Institute. "This is Khruschev and this is me, this is DeGaulle and this is me." Then he added disarmingly, "You know, of course, that I am a boaster." He was especially proud, however, when he showed me the picture of the volleyball team on which he played.

At the beginning of the fall of 1969 I was fortunate to have frequent association with Andrei Mikhailovich in Tsakhkadzor, where we went for a Rochester Conference on accelerators and where A.M. Budker presented a review talk on colliding beams. As we walked along the paths of the sports complex, he talked about the prospects in high-energy physics. Andrei Mikhailovich worried a great deal that the collisions of protons with protons or with antiprotons at very high energies could be uninteresting because the hadrons "look like mush." He thought that the main direction should be electron-positron colliding beams.

The next meeting that I remember took place in "Uzkoye."[5] Andrei Mikhailovich was not feeling very well. He was anxious about the results of the upcoming elections at the Division of Nuclear Physics of the Academy of Sciences and called to ask me to come to "Uzkoye." Our conversation was basically about the elections. Andrei Mikhailovich had a very serious attitude toward them and tried to find a reasonable balance in the struggle of different scientific groups. In 1970, before the voting, Andrei Mikhailovich delivered a very persuasive speech about the need to elect A.N. Skrinsky as a member of the U.S.S.R. Academy of Sciences. He ended his speech thus: "The only shortcoming of Alexander Nikolayevich is his youth. But, as you well know, this shortcoming will disappear when he grows older."

I remember a conversation with him in the lobby of the Moscow House of Scientists, where the meetings of the Division of Nuclear Physics usually took place (and still do). Andrei Mikhailovich enthusiastically said that he wanted to create an anti-hydrogen atom and asked

[5] Editors' note: "Uzkoye" is the name of a small sanatorium in Moscow of the Academy of Sciences.

V.N. Gribov[6] and me whether such an object would be interesting from a theoretical point of view. Our reaction was rather unenthusiastic. So he energetically scolded us.

One of my last meetings with Budker was in his small apartment in Moscow on Vavilov Street. Andrei Mikhailovich lay in bed, on a blanket, holding one hand on his chest. He complained that he had heart pain, that he was tired, that he could not struggle any more against the authorities in Akademgorodok, and then unexpectedly asked whether it was true that he might be moved with a group of scientists to work at ITEF.[7] I think he just wanted to express himself. He looked very exhausted. "Probably you think that the main thing is principle. In fact, however, the main thing is compromise. The world exists only thanks to compromise."

His eyes looking at me were full of wisdom. It was obvious that he understood much more than words could express.

---

[6] Editors' note: V.N. Gribov, theoretical physicist and corresponding member of the U.S.S.R. Academy of Sciences.

[7] Trans. note: Institute of Theoretical and Experimental Physics, in Moscow.

A Wonderful Scientific Imagination

Ya.B. Fainberg

In the beginning of the 1950's, K. D. Sinel'nikov[1] received a large paper written by a physicist whom we in Kharkov did not yet know. Having looked it over, K.D.[2] wanted me to become very familiar with it and then discuss it with him. Naturally, I could not have expected then that this work would become one of the most striking and strong impressions in science on me and that we would become close friends with its author, A.M. Budker. This study, or, to put it more precisely, this combined series of studies, showed very clearly the nature of its author — a powerful natural gift, wonderful strength, boldness and originality of ideas that were fresh and clear, like water from a spring, a very deep and keen understanding of physics, and a good knowledge of mathematical methods which played an important although subordinate role and which were fairly easy to handle because of the clarity

---

[1] Editors' note: K.D. Sinel'nikov (1901-1966) was at that time the director of the Kharkov Physico-Technical Institute.

[2] Editors' note: Kirill Dmitrievich.

and intuitiveness of the physics. The impact of this paper became even stronger because of two circumstances relevant to myself: it concerned a subject that was very close to the one on which I was working, and the author, as it soon turned out, was rather young.

Having mentioned many of Andrei Budker's typical characteristics, I still have the feeling that something, perhaps a main thing, has been missed, something that is needed to fully describe the inimitable originality of his talent. I should probably speak in particular about his wonderful scientific imagination. Soon after my acquaintance with him, it became clear that this feature was one of his strongest points, even though at first it by no means caused unanimous delight, but quite frequently irritation and scepticism. Subsequent years have proved convincingly that there was nothing more realistic than Andrei's scientific "fantasies." Nearly all of them either have been carried out by him or are being carried out by his disciples and colleagues, and they became the basis for new and very important directions in key areas of physics.

You, the reader, should pay attention to how his ideas appeared and developed: first, there appears an unexpected, extremely bold, and deep idea; however, the process of elaborating it leads to a seemingly insurmountable difficulty, perhaps even a fundamental one; then, when the main parameters are deficient by eight to ten orders of magnitude for being able to implement it, and when talented but overly cautious physicists who are always attracted to conventional solutions and who are stuffed with various dispassionate, "firm," and "unshakable" objections recommend giving it up — this is when Budker launches a decisive attack. The result would be an idea wondrous in its beauty, strength, and apparent clarity (which is properly called the clarity of wisdom), an idea that overcomes all the difficulties that had seemed insurmountable.

This two-step process was very typical. It was like this with his explanation of the crucial role of radiation in relativistic self-stabilized beams, and with the idea of collective focusing in accelerators, and with his explanation of the need to have a cold electron current moving at nearly the same speed with a proton or anti-proton current in the method of electron cooling (which nowadays marches onward triumphantly). Likewise, so it was with his ascertainment of the crucial role of relativistic effects in the gyrocon, and so on.

After the main difficulties with the idea have been overcome, there begins the development of the theory for the extremely complicated processes that are involved, which is made easier by Andrei's deep intuition and ability to partially foresee the final result. After the idea is theoretically elaborated and has hacked its way through a thicket of hot skirmishes, when all of the arguments of those who have doubts, who don't understand, or who don't want to understand are over, the next stage begins, when it is necessary to eliminate the final objections: "Yes, the physics idea or the method is original and interesting, but unfortunately it is absolutely impracticable, experimentally and technically." At this point there begins to function Andrei's unique, multi-faceted talent for finding the absolutely right solutions for experimental and technical problems, solutions that are quite unexpected, that seem to be paradoxical, but which hit the bull's eye. Then, depending on the scale and the difficulty of the problem, he would head up either a large or small group that he fostered for solving and implementing his ideas.

Of course, eventually the independence of his disciples and colleagues and their contribution to the work increases more and more; following the example of their wonderful teacher, working with him and near him, they begin to generate strong and productive ideas of their own and thus become outstanding theorists, experimentalists, and engineers. First to come to mind are A.N. Skrinsky, D.D. Ryutov, S.T. Belyaev, and B.V. Chirikov, with whose works I am familiar.

Another very strong aspect of Andrei's scientific activity was his amazing inventiveness. Even though the concept of inventiveness in science is rather ambiguous in terms of the type of activity it involves and the fields of application, different (and even opposite) points of view concerning it are well-defined. For example, some theoretical physicists, even very famous ones, share a rather snobbish opinion that can be expressed in the following way: "Inventiveness is usually the result of vague guess-work, glimmerings, and trial-and-error. The sober mind of a highly educated theoretical physicist is 'orthogonal' to this kind of inventive style, with its hunting in the dark." Although such opinions are stated in a slightly critical and ironic manner, it is quite clear that they are shared to no small extent by those who express them.

Of course, this type of inventiveness has nothing in common with true inventiveness. This was proved most convincingly by Andrei, whose works always displayed genuine inventiveness to a high degree. His inventiveness had as its basis a very deep and keen physical intuition, along with astute and clear-headed thinking, albeit extremely unconventional.

Eventually all of this became apparent. At the time, however, what happened was that I had a paper by an unknown author, a paper that was amazing on account of its power and freshness, and I very much desired to meet the author and become acquainted with him. Soon afterwards, such an opportunity occurred. I have to say, however, that my first encounter with the author gave me much less pleasure than my acquaintance with his paper. At a conference on accelerators in Moscow, Andrei presented a report on his work, most of which I already knew about. In one section of his paper, he had used the electrostatic approximation, whose validity I had doubts about. When I asked whether he had used this approximation, he sharply responded: "What fool would use the electrostatic approximation here?" His answer caused general discontent, and, as for myself, feeling insulted even though having the best of feelings for the work, which I liked, and its author, I left the room in protest. The next morning I heard someone knocking at the door of my room in the hotel where all of us from Kharkov were staying. There stood the one who had offended me. "Yesterday, Igor Vassilievich and Vladimir Iosifovich[3] gave me a good scrubbing down because they thought I had behaved very tactlessly to you," he said. "They advised me to apologize, and therefore I am doing so." While saying this, Andrei did not appear to be repentant in the least, and my feeling was that he considered his response of the previous day to be quite in order. Then he continued, smiling: "Do you know why I got mad? The thing is that I actually had used the electrostatic approximation originally, but I later redid this part." I felt that he had nothing like the intention to insult me, and we were quickly reconciled, to our mutual delight. Thus began my acquaintance and subsequent friendship with Andrei Budker.

I have described the impression that Andrei's papers made on me. A few years later I had an opportunity to observe that most of the physicists around the world who were working in the field of charged particle accelerators also accepted his work in the same spirit.

---

[3]Editors' note: Academicians I.V. Kurchatov and V.I. Veksler.

In 1956, the first International Symposium on High-Energy Accelerators was held in Geneva. This was the first meeting in this field, in general. Also, this meeting occurred after the end of the Second World War, during which scientific contacts had been interrupted, and now almost all accelerator physicists were convinced of the urgent need to resume these contacts. Finally, many new and extremely interesting results had been obtained about the physics of accelerators, and great success had been achieved in developing accelerators for relativistic particles. For all of these reasons, the symposium turned into a rushing torrent of new ideas and amazing new results about creating resonance relativistic accelerators of electrons and protons, based on the principles of autophasing that had been discovered by Vladimir Iosifovich Veksler and, independently and a little later, by E. McMillan. The strong-focusing accelerators of Christofilos, Courant, Snyder, and Livingston had already successfully started their "parade." Suffice it to recall that it was at this symposium when for the first time the opinion was expressed that implementing the idea of colliding beams could be realized. This was stated in the report of Kerst about accelerators with colliding proton beams and in the report of O'Neill about colliding electron beams. Of course, an enormous amount of work would be needed along the way in order to create such accelerators, along with suggestions on how to develop accelerators with colliding electron-positron and proton-antiproton beams, which would require fundamentally new ideas and the solution of the most complicated physical, technical, and technological problems that would arise during the implementation of these ideas. A significant, and in many cases determinative, contribution to the solution of these problems was made by Andrei Budker.

The main ideas of the methods of collective acceleration and collective focusing, which became a completely new direction in the development of accelerators, were presented at this symposium. Many innovative and interesting ideas were formulated and elaborated: among them were the idea for alternating-phase focusing in accelerators with a constant magnetic field (a circular phasotron), and the idea for an accelerator with a helical magnetic field.

In addition to V.I. Veksler and E. McMillan, the participants at the symposium included one of the creators of the first charged particle accelerators, Cockcroft; the inventor of the cyclotron, Lawrence; the inventor of the betatron, Kerst; the inventor of the principle of strong focusing, Christofilis; as well as Courant, Snyder, Livingston, Alvarez, and many others who established the basis for the physics and technology of accelerators. Also at the symposium were Adams, Sessler, Blewett, Wideröe, Walkinshaw, Mullett, Fry, and Lawson, along with Russian physicists — Vladimirsky, Sinel'nikov, Dzhelepov, Wal'ter, Kolomensky, Komar, Baldin, Naumov, and Belyaev, who were later to become (along with, of course, others who did not attend this symposium) well-known scientists in the field of accelerator physics and technology. The work done under the leadership of Minz, Rabinovich, and others was also reported, being accorded a successful reception.

I have described the Geneva Symposium in such detail because this was to be an important milestone along Andrei's scientific path and also because I wanted to give an idea of the dignified audience and the scientific authorities before whom the presentations were delivered. Unfortunately, Andrei did not himself attend the symposium. His report and one that had been done in collaboration with him were presented by Naumov. Andrei's work on the relativistic self-stabilized beam aroused great interest and highly impressed all the participants and, for many of them, was a real revelation. Budker's report and two other Soviet reports on collective methods of acceleration became one of the great events of the symposium, being discussed widely and animatedly, and thus receiving a big response. Although Vladimir Iosifovich Veksler, who came to Geneva early for a meeting of the organizing committee, tipped us off that Andrei's report and our reports had generated very great interest and that a serious discussion would occur at the symposium, the reality exceeded all of our expectations. We were truly pleased by the discussion, which continued for an hour and a half. During this discussion, deep and serious and difficult questions were mixed with what we felt were questions from talented physicists who knew little about this field of research and for whom the ideas were quite unexpected. Five or ten years later the collective methods of acceleration and focusing became quite commonplace, and many of those people who had asked questions at the Geneva Symposium became the originators of very interesting investigations in this field.

The Geneva Symposium and the preceding All-Union Soviet Accelerator Conference in 1956 were also very important for Andrei in that they let him be convinced of the correctness of his ideas and directions for research and, without a doubt, increased his self-confidence. From a scientist who was known to a small circle of people, he went to being a leading scientist, one of the most outstanding and talented physicists working in the field of accelerators.

As is well-known, there were two very important ideas in Andrei's work: one is the idea of collective focusing and the other, related to it, is the idea of the relativistic self-stabilized electron beam. In the latter, electrons confine ions and vice versa, and the process of electron radiation plays a crucial role in establishing equilibrium. To implement these ideas, it was necessary to have high-current relativistic electron beams of sufficiently large energy, with parameters that can only nowadays be obtained. Thus, it is only in recent years that interest in Andrei's idea about collective focusing has been revived and theoretical and experimental studies have been completed with encouraging results. The work of Rostoker and his colleagues on setting up a toroidal ion accelerator, with focusing provided by the space charge of electrons, deserves to be mentioned here. The results, which were obtained on modified high-current toroidal betatrons (by Rostoker, Kapetanakos, and Sprangle), are very important for implementing this method. As for the self-stabilized beam, this has not yet been created. Along the road leading to its achievement, there have been encountered technical as well as physics difficulties, related, in particular, to the long time required for establishing a stationary state, to its stability, to the problem of injection, and so on. Nevertheless, the investigations concerning the problems of equilibrium and stability and of the effect of radiation on the equilibrium of an interesting object like the self-stabilized relativistic beam led to the emergence of a line of research that is now developing very rapidly — namely, the physics of relativistic beams of charged particles and the collective processes within them. Andrei's contribution to the emergence and development of this field has undoubtedly been crucial.

During the years that followed, Andrei concentrated his efforts on implementing the method of colliding beams. Nowadays the enormous advantages of this method seem to go without saying, and attempts to call its necessity into question look naive and silly. But it was not always like this. Even after the first experimental success, it was far from being evident to everyone that the subsequent development of high energy and elementary particle physics would be absolutely impossible without the method of colliding beams. I remember the international conferences in Dubna in 1964 and in Tsakhkadzor in 1969, at which Andrei defended the concept of colliding beams with great vigor, cogency, and unassailable logic.

When he defended his ideas and the need to implement them, Andrei was far from always receiving unanimous support and a "sympathetic" attitude from certain theorists and experimentalists, and so he had to rebuff them. His responses could be sharp and even tough (although this hardness eventually disappeared); however, this was always essential and was based on clear arguments and strong, polished logic. Also very important, of course, were the extraordinary rapidity with which he responded and the unpredictability of his arguments. More than once I had occasion to witness these features during our friendly discussions. I recall how strongly impressed Christofilos was — himself an extremely talented physicist, who was also distinguished with wonderful originality and boldness of thought. I was a witness of when they met, perhaps for the first time, during a conference break at Dubna in 1964. That was when the results of the work of Andrei Budker and his colleagues on colliding beams were first reported, stimulating great interest and a vigorous discussion. The focal point of their discussion concerned the stability of the Christofilos E-layer and the effect on it of synchrotron radiation. Unfortunately, I no longer can recollect the exact arguments of both men, but I do remember how for several minutes Christofilos expressed ideas that seemed to be very convincing and beautiful, whereupon Andrei immediately refuted him with "ironclad" arguments. It made a flabbergasting impression on Christofilos. I do not know for sure, but I suspect that he had never before encountered such an opponent. Naturally, when they discussed other issues, they could exchange roles, with the other coming out on top.

Occasionally, to be sure, it was the case that the refutation he had immediately come up with, the arguments that at first had baffled everyone and caused something like "tetanus shock," were later, in the course of a more detailed consideration, found to be able to be convincingly disproved. Still, even in this case, his ideas were very unconventional, and responding to them required considerable efforts and led to a much more clarified and focused picture of the issue under discussion.

The International Conference on Accelerators held in Tsakhkadzor (1969) was of tremendous significance for the subsequent development of colliding beams and other new methods for creating accelerators with high energies. In his very convincing, clear, and impressive presentation, Andrei compared the method of colliding beams with the conventional methods and, with calm logic, defined the regions of application for each of them. While extolling the method of colliding beams, he was a bit "impolite" toward the other methods in his aggressively controversial fervor, perhaps not even noticing it himself, when he said concerning the other methods that "even the greatest optimism regarding the new methods of acceleration will not allow the attainment of the energy that can be achieved with the method of colliding beams," even though he, of course, understood better than anyone else that such a comparison was unjustified. By all means, it is necessary to develop new methods, but it is also necessary to keep in mind that they can be most effectively utilized in a system with colliding beams. After his talk, the two of us had a little conversation on this subject. In general, I was very pleased to meet with him several times in Tsakhkadzor. His talent and also his work and that of his colleagues received general acknowledgment. Also, his physical appearance was as powerful as his scientific reputation. I recall how he, with his strong, stocky build and seemingly unassailable health, swam quite fast in the swimming pool, which at the altitude of Tsakhkadzor takes quite a bit of effort. It was natural that his Herculean health seemed to be nothing to worry about.

The last time we met was in Akademgorodok, in 1974, when I attended the International Conference. Because we arrived at Novosibirsk late in the evening, I did not telephone Andrei. In the morning, when I went out to the lobby of the hotel, my first question to the person who was handling registration was, of course, about Andrei's health and whether he would be present at the opening of the conference. I did not have enough time to comprehend his surprised look when I felt someone approach me from behind, take me firmly by the shoulders, turn me toward himself, and kiss me. Everything about Andrei was strange. He had a beard; this was the first time I had seen him with it, and it emphasized his wisdom and originality, but initially made him look quite different. He seemed to manifest some kind of (I'm almost afraid to say it) tenderness. In spite of his ill health, he looked robust and cheerful. His presentation at the conference, very deep and picturesque, showed that he was in wonderful form. Andrei and his wife hospitably invited me to dinner at their home, during which he casually mentioned the new results that had been obtained in implementing the method of electron cooling. His remark elicited great interest on the part of his guests, who asked a whole stream of questions. Our conversation focused on the decisive experiments that were the turning point in the development of the electron cooling method. I remarked (in a low voice) that I was surprised he could spend so much time during dinner talking about various trifling matters (actually, here I used a more expressive word) and then mention this event only at the end, to which he laughed cheerfully and asked me to translate my remark to the other guests. Later, outside, under tall pine trees, he discussed with Drummond and me the autoresonance acceleration method of Sloan and Drummond with great interest and personal involvement. Late that night, as we walked the rather long distance from his house to the hotel where we were staying, we had a warm, cordial conversation that I remembered afterwards for a long time. It was impossible even to suppose that this would be the last time we would ever meet each other.

After this trip to Novosibirsk, there were occasional telephone calls and very warm conversations, not too often, but always giving great pleasure to me and my family.

In these brief and obviously not very coherent reminiscences of mine, I have tried to portray some of Andrei's features — his great and wonderful and unusual talent, and his originality, perhaps uniqueness, as a physicist. His work and everything he created — the very important directions in plasma physics and accelerator physics, the wonderful Institute — all have passed the most difficult test, the test of time. Year after year, as happens only with the most outstanding people, the significance of everything he did constantly increases. The large group of disciples whom he fostered, support and improve the traditions that he left. Many of them have become famous theoretical physicists, experimentalists, and engineers who have been very successful in implementing the ideas of their wonderful teacher and now their own ideas as well.

As for myself, my friendship with Andrei Budker was a great blessing.

A Creator of New Ideas

V.A. Sidorov

Our first ten years without Andrei Mikhailovich Budker — an outstanding physicist, founder of the Institute of Nuclear Physics of the Siberian Division of the U.S.S.R. Academy of Sciences, the man who played a definitive role in the lives of not only all the leading scientists of the Institute but also many scientists who now head up other scientific institutions — have come to an end.

Casting my mind back to the beginning of the 1950's, I recall my first encounter with Andrei Mikhailovich — a seminar, at which the young theorist Budker presented his fantastic idea for creating a so-called stabilized electron beam. The possibility that this type of "edifice" from electrons and ions can exist is based on relativistic effects (on the laws of Einstein's mechanics), and in our normal world of low speeds it seems to contradict common sense.

For us, young physicists who had just graduated from university, it was difficult (due to the regulations in those days) to attend the seminars of a neighboring department in the Institute, whose name is now the I.V. Kurchatov Institute of Atomic Energy. However, the rumor that this seminar promised to be very interesting gave us enthusiasm, which was to be rewarded.

Following his brilliant report there occurred a curious incident (which adds something to the portrait of Budker at that time). Answering many questions and feeling "up in the clouds" due to all the attention from prominent persons, at the same time quite exhausted, the speaker, instead of responding to a whole series of questions asked by academician V.I. Veksler, declared that "sometimes even a hundred wise men cannot answer every question." For this pun, Budker received a strict scolding from academician L.A. Artsimovich, who presided over the seminar, and was disliked for many years by the offended Veksler. However, I must point out that it was precisely academician V.I. Veksler who visited Novosibirsk in 1964 and played a decisive role in persuading members of the Division of Nuclear Physics of the Academy of Sciences to elect Andrei Mikhailovich as a member of the Academy.

My first direct contact with Andrei Mikhailovich took place in 1961, at the time when Budker's former Laboratory for New Acceleration Methods, which had already been transformed into the Institute of Nuclear Physics of the Siberian Division of the U.S.S.R. Academy of Sciences, was preparing to move to Siberia. The head of that part of the laboratory that was to remain in Moscow, B.G. Yerozolimsky, suggested that I come and work in his group. After becoming acquainted with the situation, I stated in a conversation with Andrei Mikhailovich that I could not be interested in the work of the rear guard of his Institute, because everything promising was moving to Siberia. An invitation to head up one of the main laboratories of the Siberian institute was immediately made, in violation of a previous agreement between Budker and Yerozolimsky. As he fervently urged me to move to Siberia, one thing Andrei Mikhailovich said was that if I refused, he would be forced to appoint Sasha Skrinsky, who had been only a student until almost the day before, as the head of this laboratory. Incidentally, Skrinsky did become the head of a laboratory just one year after we both simultaneously moved to Siberia.

In the initial years of his directorship, during the time when the Institute was taking shape, Andrei Mikhailovich, in order to justify the need for an organizational measure, often appealed to the fact that he, a theorist, had already long ago constructed a theory for the Institute's organization and that what was now transpiring was an experimental check of this theory.

"Everything that we have already managed to do has not been fortuitous; everything corresponds to a previously constructed model, so trust me that I also know what I am doing now."

I am not convinced that this was always true, but there is no doubt that he was always interested in sociological issues, the structure of society, and social policy. Of course, this was the case even before he entered upon the post of director, as the leader of a large organization. Andrei Mikhailovich told us that many years before becoming a director, he had constructed for a collective farm a theory that would surely raise our country's agriculture to a high level. In his theoretical model, it would not be the collective farm, but rather each of its members that would be a millionaire.

In his "sermons" that he addressed to his closest colleagues — and this term is quite appropriate for his frequently moralistic reasonings — Andrei Mikhailovich combined in a wonderful way two seemingly contradictory principles of scientific leadership: the collective and open method of making decisions, and the particular role of a talented individual, a creator of new ideas. According to his first principle, the director, who is the leader of the scientific organization, must be surrounded by his scientific colleagues and not by a powerful administrative staff.

"Otherwise," said Budker, "the administrative staff, by gathering closely around the director and looking to him with 'love' and with hope for his favor, will turn their backs on the scientific staff."

Andrei Mikhailovich was constantly surrounded by scientists, with several discussions going on in his office at the same time; I was always amazed at his ability to work and think in an atmosphere of incessant din and his tolerant attitude.

However, the main weapon for "glasnost,"[1] the weapon invented by Andrei Mikhailovich for collective leadership, was the Round Table — the daily meetings of the leading members of the Institute that were held in a free and easy atmosphere.

---

[1] Trans. note: "Glasnost," which has become a well-known concept in recent years, refers to a situation of apparent openness.

Three generations: left to right, V. E. Balakin, Lenin Komsomol Prize winner and Candidate in Physics and Mathematical Sciences; V. A. Sidorov, corresponding member of the USSR Academy of Sciences; and academician A. M. Budker (1972).

The number of these members increased, and the "Knights of the Round Table" had to be reorganized: for many years now, the members of the Scientific Council of the Institute have been meeting weekly on Wednesdays, and on the other days the various topical sections of the Council meet at the Round Table.

Everyday at 12 o'clock noon, people gather for a cup of coffee; the participants are those persons who had been invited by Andrei Mikhailovich, and later also by his disciples, solely on the basis of one criterion — talent for doing research in physics.

"So few people have this particular talent," Budker used to say, "that if you take other criteria into account — character, background, education, even intelligence in the usual sense of this word — you will necessarily narrow the possible choices so much that you will collect bad physicists. Nevertheless, however, there is one more characteristic that is necessary — this is honesty."

While giving due significance to the issues related to organizing a group, Andrei Mikhailovich always emphasized that in science, as in no other field of activity, the role of a creative individual is very important. In science, especially in fundamental research investigations, the personality of the scientific leader plays a determinative role. This is the person who formulates the research problem and in whose mind there takes shape an idea for ways to get to the goal. A scientific group — big or small, institute or laboratory — must be created around its leader.

Andrei Mikhailovich often stated the following rule, to which he firmly adhered: If the head of a laboratory leaves, the laboratory disbands. Throughout the history of the Institute, it never happened that there was a successor to a laboratory head who left, nor was a laboratory that remained without a leader ever assigned a new head.

One exception to this rule, however, was made — for the entire Institute. Following the death of Andrei Mikhailovich Budker, his favorite disciple — A.N. Skrinsky — became the head of the Institute. But Budker was an exception himself. He did more than merely create a scientific school. From "under his wing" came out approximately ten directors of institutes, leaders of large projects. An exception, as Andrei Mikhailovich liked to say, only confirms a rule.

For 20 years now, the Institute of Nuclear Physics has been carrying out an economics experiment, whose goals, with surprising exactness, coincide with the program of "perestroika."[2] With its powerful facilities for pilot production, the Institute not only develops, but also delivers to enterprises in the national economy the most modern technological equipment, amounting to about ten million rubles' worth annually. Interestingly, the rules and policies for this activity were composed by Andrei Mikhailovich, who had a rather poor idea of the cost and magnitude of work to be expended on specific developments. He almost always made gross errors in these estimates, and always in the same direction — lowering them by large factors. It can be said that he was an intuitive economist, keenly sensing the general laws of economics, deducing them not from specific calculations but from certain philosophical considerations — for example, that one must measure the quantity of milk not per cow, but per milkmaid.

Almost ten years of our lives have passed without Andrei Mikhailovich Budker, but it is precisely now, during the grandiose reconstruction

---

[2] Trans. note: "Perestroika" refers to reconstruction in society.

of all of Soviet society in accordance with the principle of "glasnost," with priority on the pulling of economics-related levers, that the ideas and words and deeds of the creator and director of the Institute of Nuclear Physics, academician Andrei Mikhailovich Budker, stand out most clearly in our memory.

Fearless Innovator

W.K.H. Panofsky

I met Professor Budker first in 1956 when the Soviet accelerator and high-energy physics facilities were opened up for the first time after World War II to visitors from the United States. This was indeed a memorable visit, but the highlight was my meeting Professor Budker for the first time in Moscow at the Atomic Energy Institute. He showed us personally various high-current devices of novel design. Of particular interest to me were his methods of shaping boundaries of magnetic field without the use of iron by using copper sheets so that the eddy currents in a changing magnetic field would produce the necessary field configuration. He impressed the visitors, including myself, with this elucidation of a very high-current betatron; at the time none of us had the quickness of mind to analyze that proposal in sufficient detail to discuss it critically. We saw many other devices and I started discussions with Professor Budker there and began a friendship which we maintained for the rest of his life. I was persuaded from the beginning that he was indeed a great inventive genius.

At a future occasion in Moscow Professor Budker and I had discussions about the possibility of collaborating on building a joint storage ring laboratory as a bilateral undertaking linking our institutes, since SLAC and the Novosibirsk Institute had so many common interests. Unfortunately, this idea, as many other inventions of Professor Budker's, was slightly ahead of its time. Both of us felt that common interests in science might be one of the healthiest reasons for bridging political differences. I still hope that will become a reality.

Let me digress to an amusing incident: Professor Budker visited our laboratory once and he decided to bring a gift to me and my wife. Since Professor Budker always "thought big" he decided to bring a Siberian bear skin, to be made into a rug. It was not cured as yet. He told me how the rug was brought through the American Customs: he said the Customs officer opened the bag and the bear skin stank so badly that the official closed the bag rapidly without examining what was in it. So therefore no customs duty was due. We tanned the skin properly and made it into a rug and it now graces our house in memory of Gersh Itskovich Budker.

My last visit with Budker was at his home in Novosibirsk, after he had just recovered from his previous heart attack. We talked about many new ideas for acceleration. He repeated his appeal that one should not just make accelerators bigger but that new technologies were essential, and we talked about the prospects for achieving controlled thermonuclear reactions. His son, then 11 years old, was doing the translating! I left that meeting very depressed, considering Budker's state of health; he had so many ideas which needed elaboration and all we can do is to do our best to live up to the standards he set.

Introductory Talk to the G.I. Budker Memorial Session[1]

On April 25, 1978, a Memorial Session was held at the Washington meeting of the American Physical Society in memory of Gersh (Andre) Budker, who died July 4, 1977. Papers were given at that session by four American physicists whose work closely touched on and benefited from Budker's work. A telegram was sent the previous day by Professor A.N.

---

[1]Published in the book *Problems of High-Energy Physics and Controlled Thermonuclear Fusion*, edited by S.T. Belyaev (Nauka, Moscow, 1981), pp. 9-10.

Skrinsky, Chairman of the Organizing Committee of a seminar held in Novosibirsk, as follows:

> Please telephone to Professor Panofsky in Washington the following on behalf of colleagues and friends of Professor Budker gathering here in Novosibirsk at the seminar devoted to the sixtieth anniversary of his birthday. We warmly greet our colleagues and friends in America participating in the Budker Memorial Session.
>
> Brilliant physicist, remarkable and attractive personality — he also made great contributions to international collaboration of physicists. We believe that the ideas of Professor Budker will inspire and stimulate our mutual activity for many years.

> Sincerely,
> Skrinsky
> Chairman, Organizing Committee

Professor Norman Ramsey, President of the American Physical Society, and I, as Chairman of the Memorial Session, replied as follows:

> We deeply appreciate your greetings to Budker Memorial Session of the American Physical Society. We reciprocate greetings to your birthday and memorial gathering. We share your admiration for Professor Budker's scientific contributions and expect them to provide the basis for major future advances in physics.
>
> May this dual recognition of Budker be a symbol of our continuing close relations.

> Norman Ramsey
> President, American Physical Society
>
> Wolfgang Panofsky
> Chairman, Memorial Session

It is indeed a fact that Professor Budker 's work has been an important element in the commonality of interest between physicists in the United States and the Soviet Union. When in 1956 it was first possible for American physicists to visit particle accelerator institutions in the

Soviet Union, the work of Budker, then at the Kurchatov Institute, drew immediate attention. Many of the ideas originating from that visit led to extensive studies in the West and this cross-fertilization of ideas continued after Budker transferred his work to the Institute of Nuclear Physics of the Siberian Branch of the Soviet Academy of Sciences. The papers contributed by the speakers at the Budker Memorial Session dealt with the main lines of Budker's work at Novosibirsk. He developed novel accelerator ideas with particular emphasis on storage rings. He innovated in theory and practice of controlled thermonuclear reactions and he developed high-powered pulsed devices of many kinds. He continued to explore the basic limitations of accelerating and storage ring devices and his many inventions were aimed at pushing back those frontiers. In particular, his concept of electron cooling opens new opportunities for the construction of storage rings for protons and antiprotons whose large mass prevents shrinking of the radial phase space through radiation, as is the case for electrons.

Many of Budker's contributions were ahead of their time in the sense that he did not live to see their full impact on science and technology. It is therefore left to the next generation of physicists in all countries to pursue these ideas. I hope that the papers presented at the Budker Memorial Symposium give an indication how Budker's influence has been felt in developments in the United States and how this work might evolve in the future.

Obituary for Gersh I. Budker[2]

Gersh Itskovich (Andre) Budker died July 4, 1977, after an extraordinarily productive career. His work, characterized by an unfettered ingenuity and skill, is well-known to his many friends throughout the world working on accelerators and plasma physics.

Budker was born 1 May 1918, and graduated in 1941 from the University of Moscow. In 1946 he joined Igor V. Kurchatov at the Institute of Atomic Energy of the U.S.S.R. Academy of Sciences and worked on many phases of atomic energy, including the theory of graphite-moderated reactors. In 1956 he became a professor at the Moscow Engineering Physics Institute.

[2]Published in *Physics Today*, September, 1979, pp. 78-79.

I remember meeting Budker at the Kurchatov Institute in 1956 during the memorable first visit by American high-energy physicists to various particle-accelerator institutions in the Soviet Union. At that time he showed us a number of advanced devices whose promise appeared without precedent in Western accelerator technology. There was an ironless betatron, and magnetic-beam transport channels with boundaries set by eddy currents, rather than magnetic-pole faces. There was a space-charge-stabilized model of a high-energy, high-current electron circular accelerator. All these devices led to intensive critical studies at home. Shortly after this visit, Budker transferred his work to a new institute of the Siberian branch of the Soviet Academy of Sciences. The facility, of which he became director, is the Nuclear Physics Institute and is located in the "academic city" near Novosibirsk. In 1964 Budker became a member of the Soviet Academy of Sciences.

Budker's work in Novosibirsk continued to concentrate on the development of novel accelerator ideas, controlled thermonuclear reactions, and pulsed high-powered devices of various kinds. He shunned the mainstream of Soviet (and also Western) accelerator design — his disdain of scaling up established accelerator designs to larger and larger powers was well-known.

In general, his developments were frequently ahead of their time, and he always insisted that completed systems were constructed of components designed and built according to the most advanced principles at his institute, or with the help of surrounding industry. All this added up to a pattern of intensive creativity in a certain degree of isolation. As a result, Budker's developments were always challenging technologically, but, not infrequently, final success was elusive since the performance of some advanced component fell short of expectation. In many respects it was just this pattern that made a visit with Budker at his institute a memorable experience. New ideas were constantly being introduced and many of these came to fruition, either at his institute or abroad.

A substantial part of the work of Budker's institute concentrated on a family of electron-positron colliding-beam storage rings, which were designated by a series of numbers from VEPP-1 to VEPP-4. Two of these installations became productive as tools in particle physics.

A small group of elementary-particle physicists exploited these machines, in particular VEPP-2 and its high luminosity version, VEPP-2M. These storage rings made important contributions to elementary particle physics, in particular by determining the precise masses and decay widths of vector mesons and the branching ratios of their decays. Budker gathered around him a group of able theorists who made important contributions in many phases of accelerator and plasma theory including various collective-instability phenomena and the theory of polarization of stored electron beams.

Budker's storage ring VEPP-3 is being used for synchrotron-radiation research, and he had been planning to supply storage rings for this new field of endeavor to other institutes in the Soviet Union. VEPP-4 is nearing completion and is expected to reach higher $e^- - e^+$ collision energy — 14 GeV — than any attained to date. One of the most fruitful accelerator innovations that Budker originated was the idea of electron cooling. This is the method by which the phase space of protons stacked in a storage ring is shrunk by having the protons interact with a stream of electrons traveling at the same velocity. The electrons exchange transverse momenta with the protons. This idea was demonstrated experimentally at Novosibirsk and is now being exploited intensively both at CERN and at Fermi National Accelerator Laboratory for the construction of antiproton storage rings desirable for the achievement of antiproton-proton collisions. This work, if successful, would be a memorial to only one of Budker's many creative ideas.

Not only did Budker innovate technologically, but he also introduced novel social patterns of operating a creative laboratory. Part of the effort of his institute was dedicated to the design and production of low-energy accelerators, including pulse-transformer accelerators, and of high-powered microwave tubes that could be marketed both at Soviet institutes and medical centers and also abroad. From the proceeds of these sales, Budker was able to support his program more flexibly than would have been possible through exclusive dependence on government support.

Budker was a strong exponent of increased collaboration with the West in the accelerator arts and particle physics. In this field, as in the science itself, his ideas and plans were frequently ahead of his time. His initiatives for major joint undertakings in which Western technology, particularly in data processing, would complement some of the achievements of Budker's laboratory, have not yet come to fruition. No better memorial to Budker's work could be made than a practical realization of his dream of collaboration in storage-ring physics bridging Siberia and America.

Budker was an exceedingly capable analyst and designer as well as a fearless innovator. He surrounded himself with a young group of associates and students who shared with him responsibility for major decisions at the famous "round table" council. His disciples will perpetuate much of his style and ideas, but a great driving spirit is gone.

A Sight to Behold

V.N. Bayer

In April of 1955 I came to "settle" in Moscow. I had letters of recommendation to some physicists, including S.M. Rytov. He sent me to the Laboratory of Measurement Instruments (LIPAN) and, in particular, to A.M. Budker. Andrei Mikhailovich invited me over to his home. I had to go by tram quite a ways, to the outskirts of Moscow. A thick-set, energetic, bald-headed man with traces of red hair met me at the door. The room we went into was almost empty. It contained only a shabby desk and two chairs. Budker asked me about my studies in Kiev, about the educational system at Kiev University, and about the students' scientific discussions, during which I had gotten into "hot water": I had praised Paul Dirac, whereas my teachers in Kiev had criticized him as an abysmal idealist. "Why so?" said Andrei Mikhailovich in surprise. "Dirac was the one who created the theory of representations for quantum mechanics, from which it follows that the physics conclusions of a theory are independent of the choice of basis, and consequently the arbitrariness in the choice has no influence on the results.

Therefore, it was precisely Dirac who made a great contribution to the materialistic interpretation of quantum mechanics."[1]

During our conversation, most of the time was spent solving physics problems. Nowadays these problems are assigned to students at Novosibirsk State University, but at that time, this "test of a gentleman" was given to everyone who applied to enter Budker's group. "Unfortunately," he said at the end of the conversation, "doing the paperwork takes about one year in our organization. So, if you don't find a position elsewhere, put in your application here and then return to Kiev and await the results."

My second encounter with Andrei Mikhailovich occurred in November of 1958, after I had already finished my post-graduate studies. He now lived in another apartment — near the "Sokol" subway station. I remember that we talked in a large room with many windows. After some questions about my post-graduate activities, Budker with great enthusiasm began to explain the idea of colliding electron beams. His opinion was that colliding beams would become the main program of the Institute of Nuclear Physics, to be established in Novosibirsk. "If we are able to ram 100 amperes against 100 amperes, we will have no problems at all." The 100 ampere current was the major topic of our conversation; this would eliminate all difficulties. Once again I was tested by having to solve problems, and, as a result, I received an offer to work at the Institute of Nuclear Physics of the Siberian Division of the U.S.S.R. Academy of Sciences. At that time the Institute and the LNMU (Laboratory of New Acceleration Methods, which was also headed by Andrei Mikhailovich) were both located inside what is now the Kurchatov Institute of Atomic Energy, mainly in a building that had formerly been a medical clinic but also occupying part of a new laboratory building. It was in the former medical clinic where the first colliding beam device, VEP-1, was built.

After being enlisted into the Institute, I associated with Budker rather frequently: there were very few scientists in the group, and his inexhaustible and boundless imagination needed an outlet. Here are some episodes during the Moscow period.

At the same time that the VEP-1 storage ring and its injector were being built, a program of physics research was taking shape. Andrei

---

[1]Trans. note: At that time, scientists in Western capitalistic countries were considered to be adherents of philosophical idealism — as opposed to materialistic Marxism.

Mikhailovich wanted the attention of the country's leading scientists
to be attracted to the work of the Institute. In October of 1959, I.Ya.
Pomeranchuk came to visit us.[2] We had a very lengthy discussion,
alternating between theoretical considerations about checking quantum
electrodynamics at small distances and technical details about electron-
electron colliding beams. I.Ya. Pomeranchuk was not very happy with
what he heard. After saying good-bye to him, Budker came to see
me, complaining that for some reason our program had not impressed
Pomeranchuk and that we needed to put more thought into improving
it. My response was that the program would be much broader if we
created electron-positron colliding beams instead of electron-electron
beams. During the ensuing discussion, we exchanged sides many times,
first one of us persuading the other of their feasibility and the other
expressing doubts about it and then vice versa. The upshot was that
Andrei Mikhailovich asked me to lay aside everything else and evaluate
whether it would be possible to implement the new version. The next
day there began a period of intense work, with exhausting discussions
lasting from morning to evening. But, in something like five days, we
very roughly sketched out the VEPP-2 project. Its maximum energy
(700 MeV) was chosen so that K-mesons would appear.

Budker watched our work discipline very strictly. Because I took my
daughter to nursery school, even with the best transportation schedule
I would arrive at work at five minutes past nine. Sometimes Andrei
Mikhailovich waited specially for me, to give me a long scolding for
being late.

In the summer of 1959 I was planning to go to an alpinist camp in
the Caucasus mountains. However, just at that time, it turned out that
some calculations for the problem of the field structure in an ironless
accelerator (this type of device was very popular at the INP in the
1950's and 1960's) were not finished. Budker called me on the day of
departure and "persuaded" me to cancel my trip in one hour. The last
name on my camp permit was changed to Bayer*anov*, so that someone
else, Baranov, was able to go instead.

In the spring of 1961 we played a rather crude prank. A notice was
posted on all the bulletin boards at the Institute of Atomic Energy
announcing that on April 1, in the conference hall, there would be

[2]Editors' note: I.Ya. Pomeranchuk (1913-1966) was an academician, a two-time winner of the
U.S.S.R. State Prize, and the founder of a school in theoretical physics.

a lecture by A.M. Budker entitled "Mass Production of Accelerators in Siberia and the Tasks of the Institute for Nuclear Physics." A few minutes after four o'clock, someone called Andrei Mikhailovich and told him that more than 200 people were waiting for him in the lecture room and he was late. For a while he did not understand what was going on. He had in his office at the time some high-ranking visitors, who began to chuckle. He had to go to the lecture. According to gossip from eye-witnesses, he managed to acquit himself honorably. But the next day he undertook an investigation, having picked out some "suspicious" and "semi-suspicious" persons. Andrei Mikhailovich approached each person who was under suspicion and said, "Confess; I won't do anything to you." No one confessed.

Here are some episodes from the Novosibirsk period.

In the spring of 1962, Andrei Mikhailovich tried to draw the attention of the theorists at the Institute to the possibility of polarizing electrons in a storage ring. A year later this problem was actually solved at Moscow State University, and it turned out that polarization was a real effect. Somehow Budker had sensed this effect without any calculations.

In September of 1963, the first foreign physicists visited the INP; among them were B. Richter, B. Gittelman, and F. Mills. After the visitors had been shown the Institute, it was decided to take them for a ride on a cutter. A hard autumn storm came up at sea, and the cutter pitched wildly, so the captain decided to go to Berdsk Bay. We were unable to have dinner in the ship's stateroom until we got there. During the dinner, Andrei Mikhailovich commented that only small groups were occupied with colliding beams, all basically young physicists from the U.S.S.R., the U.S.A., and Italy, but that the future of high-energy physics lies in the direction, and that we needed to support each other both in the development and improvement of the colliding beam method as well as in organizational aspects.

At the end of 1965, Valery Katkov (who was then a graduate student) and I both became interested in the birth of neutrino pairs by high-energy electrons. It was clear that the probability of this process was proportional to the energy with a large exponent (which is typical for weak interaction processes). We thought that the effect could turn out to be important and even find practical applications if we could get to sufficiently high energy. However, our reasoning was incorrect.

When I discussed this question with Andrei Mikhailovich, he noticed that the effect must depend on the magnetic field in the electron rest frame (an invariant quantity), that is, the result involves the product of the energy and the magnetic field. The dimensionless combination containing this quantity is very small and enters the probability with a large exponent, so the effect turned out to be very weak. Budker liked to recall with pride that Landau had called him a "relativity engineer." During our discussion, this characterization was very clearly confirmed.

Frequently, at the Scientific Council of the Institute or just during various discussions, Andrei Mikhailovich appealed for unconventional thinking and for looking for unconventional solutions. The following was one of them. When Moscovites moved to Akademgorodok, it was very popular to buy a sailboat. Even in this matter Budker acted unconventionally: from two "Kazanka" boats he built a catamaran with a high deck. On the rare occasions when Andrei Mikhailovich managed to put out to sea in this strange contraption, he usually stood on the deck in his swim suit, legs wide apart like a "sea wolf," looking through field glasses. It was a sight to behold.

Andrei Budker's father.

Budker (on the right) during the Great Patriotric War: the Moscow State University graduate became an artillery officer.

The "Assembly Shop" building at the Kurchatov Institute that housed the first Russian nuclear reactor.

Bust of I. V. Kurchatov in front of the main entrance to the Kurchatov Institute of Atomic Energy.

During the first years of the Institute, he also had to be an architect (1962).

Main building of the Budker Institute of Nuclear Physics.

With discussions and arguments he was in his element.

With M. V. Keldysh, president of the USSR Academy of Sciences, at the Institute construction site (1961).

"You see, Piotr Leonidovich, there is an interesting paradox!" Academician P. L. Kapitza, one of the patriarchs of Soviet physics (on the left), and A. M. Budker, one of the youngest directors of an institute (1961).

Perhaps a new idea is being born? A conversation with V. P. Dzhelepov, corresponding member of the USSR Academy of Sciences (1961).

Even a language barrier does not stop him: Prof. Glen Seaborg, chairman of the US Atomic Energy Committee, visiting the Institute (1971).

The Institute's first-born: the VEP-1 colliding-beam device (1963).

In the VEPP-2 control room: this facility became a second university for several generations of physicists at the Institute. With Budker (from left to right) are V. A. Sidorov, I. Ya. Protopopov, S. G. Popov, A. N. Skrinsky, and V. V. Petrov (1964).

The president of the Academy is pleased: A. M. Budker with M. V. Keldysh by the VEPP-2 device. Positrons will soon be here (1964).

VEPP-2, the world's first electron-positron storage ring with colliding beams (1965).

"One head is good...

... but two heads are better": thinking about the future of colliding beams. On the left, A. N. Skrinsky (at the time a corresponding member of the USSR Academy of Sciences) (1969).

A "5" in physics: the 1967 Lenin Prize winners at the Round Table. From left to right: V. A. Sidorov, A. N. Skrinsky, A. M. Budker, A. A. Naumov, and V. S. Panasyuk.

Another "5" in physics. In November of 1968, five scientists from the Institute were elected members of the USSR Academy of Sciences. From left to right: corresponding member V. A. Sidorov; academicians R. Z. Sagdeev, A. M. Budker, and M. A. Lavrentyev; corresponding member R. I. Soloukhin; and academician S. T. Belyaev. Corresponding member A. N. Skrinsky is at the podium (not in the photograph).

Life, I love you!

A serenade (1967).

Giving an examination on the ski trail (1961).

A 1967-model handstand.

"Cook it a little bit more..." (1967).

At a final defense there were usually two protagonists: the thesis defendant and Andrei Budker.

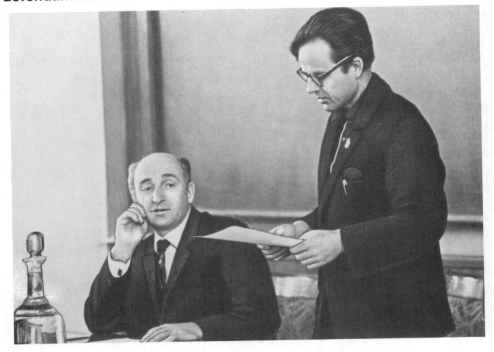

"Boring formalities." On the right, the scientific secretary of the Council, B. V. Chirikov (at that time a Candidate in Physical and Mathematic Sciences) (1966).

A rare picture: Budker in the role of listener (1964).

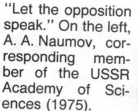

"Let the opposition speak." On the left, A. A. Naumov, corresponding member of the USSR Academy of Sciences (1975).

"Now I will explain what the thesis defendant meant." On the right, L. M. Barkov (at that time a corresponding member of the USSR Academy of Sciences) (1975).

Secret balloting (1976).

The Round Table—Andrei Budker's favorite "child" (1968).

Mechanism for seeking solutions at the Round Table:

From general scepticism...

... through persuasion...

... to concensus. On the left, A. A. Nezhevenko; on the right, G. A. Blinov (1975).

The director's "veto" (1974).

Amongst presidents.

With Charles de Gaulle, president of France (1966).

With George Pompidou, president of France, in front of the Institute (1970).

In the middle, Philipp Handler, president of the US National Academy of Science, visiting the Institute. On the right, academician G. I. Marchuk, later the president of the USSR Academy of Sciences (1973).

Budker giving explanations to Josip Broz Tito, president of Yugoslavia (1968).

M. A. Lavrentyev, president of the Siberian Division of the USSR Academy of Sciences, at the Round Table, after presenting Budker with a Certificate of Discovery (1971).

Our physics must be an advanced physics.

Andrei Budker (1973).

Budker acquaints the president of the USSR Academy of Sciences, A. P. Alexandrov, with the prospects for the colliding beam method (1977).

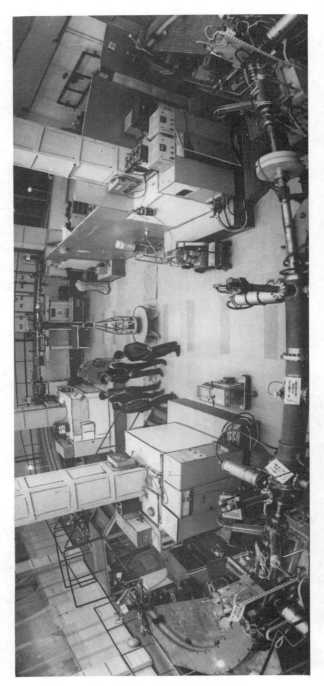

Electron cooling in action: the NAP device, in which the possibility of cooling a heavy particle beam was demonstrated for the first time in the world (1974).

Construction of the VEPP-4 facility was a significant step in the development of high energy physics in the USSR (1974).

Physics should be thrifty...

In conversation with academician A. G. Aganbegyan, an economist (1973).

Alexei N. Kosygin (Soviet prime minister), a visitor to the Institute, was interested in opportunities for using accelerators for the development of new technologies (1969).

... and useful for the national economy.

Accelerators from the Institute are widely used in industry to irradiate manufactured cables.

At the granary in the port of Odessa, accelerators from the Institute are used used to disinsectize grain.

Contemporary classics.

From left to right: Prof. V. F. Weisskopf (USA), Prof. Yu.
B. Rumer, and academician A. M. Budker (1970).

Not only scientists visited the Institute: on Budker's immediate left, Konstantin Simonov, a famous Russian author (1972).

The patriarchs of colliding beams: Andrei Budker and Wolfgang Panofsky were friends for many years (1975).

Reminiscing about the past: Budker's colleague in the 1950's Prof. B. G. Yerozolimsky (May, 1977).

Thinking about the future: with students from the Novosibirsk Physics and Mathematics School in the VEPP-4 control room.

The last photographs of Budker (June 23, 1977).

With A. P. Alexandrov, president of the USSR Academy of Sciences.

At the entrance to the Institute.

Life goes on... (1987).

"Sir Director"
The Novosibirsk Period (1958–1977)

In his sixth year as the director (1963).

Through Discussions — to a Decision

G.I. Dimov

I became acquainted with Andrei Mikhailovich at the end of the 1950's, when the physics and technology of charged particle accelerators were being actively promoted and studies of controlled thermonuclear fusion were also just taking off. At the first accelerator conference in which many international scientists participated, which was held in 1956 at the Lebedev Physics Institute of the U.S.S.R. Academy of Sciences, A.M. Budker and A.A. Naumov presented the results of their work on the stabilized circulating electron beam and on the ironless synchrotron. These investigations caused extremely high interest, due to their novelty and the boldness of the proposed solutions. During discussions, however, grave doubts concerning the possibility of successfully carrying out the proposed accelerator schemes were also expressed. Already prior to this conference, I had visited A.M. Budker's laboratory in LIPAN.[1] This group of young scientists had strongly impressed

[1] Editors' note: LIPAN stands for Laboratory of Measurement Instruments of the Soviet Academy of Sciences. (This was the name of the Kurchatov Institute of Atomic Energy in the 1950's.)

me with their creative atmosphere and enormous capacity for work. The most amazing things were Budker's inexhaustible inventiveness in accelerator techniques and the large number of new physics ideas.

However, I really came to know Andrei Mikhailovich from the middle of the 1960's when I began to work under his leadership at the Institute of Nuclear Physics, Siberian Division of the U.S.S.R. Academy of Sciences. He was the heart and soul of the scientific group. During the 1960's he carried on continuous discussions with his collaborators on practically all of the issues related to the construction of experimental facilities and the carrying out of experiments. Frequently, new solutions were conceived during these discussions. After meticulous analysis, most of the solutions were rejected for various reasons; however Andrei Mikhailovich kept on coming up with new ideas. Significantly, everything was carefully evaluated numerically. A.M. Budker was a virtuoso in this regard. He found the easiest, most physically transparent methods of calculation for the most difficult problems. His indefatigable ability to analyze physics and technological situations, repeatedly and from many aspects, was amazing. Time and again we would return to questions that had been discussed months and even years before. In my opinion, it was precisely thanks to all of this that the remarkable proposal of A.M. Budker for electron cooling of protons and antiprotons in circular accelerators was conceived. By having been a participant for many years in lengthy and substantial discussions about the problems of accelerators and controlled thermonuclear fusion, I learned a great deal. Precisely for this reason, above all other reasons, I consider myself to be a disciple of Andrei Mikhailovich.

During the 1970's, the Institute grew considerably. Andrei Mikhailovich was no longer able to involve every staff scientist in discussions, as before, although his passion for this style of work did not alter. When I was in Moscow at the same time he was, he usually invited me to go for a walk after work. After beginning the conversation with an anecdote, Andrei Mikhailovich would start to discuss the next version of VLEPP. At that time he was captivated by the notion of storing energy in a high-power circulating proton beam in order to accelerate electrons. In general, Andrei Mikhailovich was not uninterested in energy storage issues. Moreover, relativistic proton beams promised new opportunities for the fast transportation and distribution of energy. Here, Andrei

Mikhailovich's imagination, substantiated by calculations, was boundless.

In spite of his keenness for new ideas, A.M. Budker was very rational when it came to organizing experimental work. He was one of the first in the Soviet Union to perceive the enormous opportunities of colliding beams in high-energy physics and to concentrate his main efforts on the creation of accelerators with colliding beams, though at a cost to the development of his own pet conception, the stabilized electron beam. Also, he forwent elegant schemes for accelerators for radiation technology, in favor of simple ones that were more reliable. He appealed to the international community of plasma physicists, not to limit themselves to the physics of plasmas only, but actively to pursue working out the physics basis for thermonuclear fusion reactors.

Andrei Mikhailovich and Our Institute

A.N. Skrinsky

I first heard the name of Budker in 1956. Having finished my third year in the physics department of Moscow State University, I needed to ponder where to go afterwards for the pre-thesis training and thesis work and where to work after graduating from the university. And then, through some mutual acquaintances, one of the leading scientists at I.V. Kurchatov's LIPAN[1] (which was soon renamed the Institute of Atomic Energy) recommended that I join a new, very lively and interesting laboratory organized by Andrei Mikhailovich, who was then already achieving recognition. The next time I heard about Budker was in the spring of 1957 when Vadim Volosov, who at that time was already working in Budker's laboratory, unexpectedly contacted me in our dormitory. He told me about various approaches and ideas that were being developed by the young group, most interesting of which for me were the new approaches for accelerating particles to high energy.

[1] Trans. note: LIPAN is the acronym for Laboratory for Measurement Instruments of the Academy of Sciences (the original name for the Institute of Atomic Energy).

171

Vadim pushed me to make a decision about where to do my training work (I was also a little acquainted with the Nuclear Physics Institute of Moscow State University and with Dubna, although at that time I had never visited the latter). In August, after passing an interview that was not very successful — I was unable to grasp how charged particles move ("drift") in crossed electric and magnetic fields —, I was admitted into Andrei Mikhailovich's laboratory, to the group headed by Borya Chirikov. I began to work under Volosov on an experimental investigation of the phenomenon of a virtual cathode in high-current electron beams in a longitudinal magnetic field (Andrei Mikhailovich and his colleagues hoped to use this for focusing high-current ion beams in cyclic accelerators).

During my first months of work, I saw Andrei Mikhailovich rather seldom and practically never interacted with him. It was only at the end of 1957, apparently at Chirikov's recommendation, that Budker suggested I join the group that, at his initiative, went on to solve the problem of making a device with colliding electron-electron beams. From this time on for practically 20 years, I was fortunate to be one of the close colleagues of Andrei Mikhailovich.

The use of colliding beams as a practical means to achieve ultra-high interaction energies had only just become a subject of discussion in the middle of the 1950's, and by far most physicists considered it an issue of the indefinitely distant future. Yet, in many laboratories around the world, including the Institute of Nuclear Physics, people became enthusiastic about this method. Specifically, the thought was that the first step would be electron-electron colliding beams: first, because with these low-mass particles, a modest energy of about 100 MeV is sufficient for the advantages of the method to be clearly exhibited; second, because the utilization of radiative cooling, recently understood, would make it possible to store small cross-section beams of the required intensity. In addition, just at that time it was learned that the interaction law for electron-proton elastic collisions is different from the Coulomb interaction of point charges, and confirmation was needed to show that the finite size of the proton is responsible for this (i.e., to prove that quantum electrodynamics is valid on the small distances corresponding to colliding beams with energies in the hundred MeV range).

Many laboratories (including some Soviet groups) took up the problem of colliding beams, but final success — the carrying out of the experiments on electron-electron scattering — was achieved only by two research centers, and simultaneously at that: Stanford University (U.S.A.) and the Institute of Nuclear Physics in Novosibirsk, which had been organized on the basis of Budker's laboratory.

However, this did not happen until 1965, after many years that were, I feel, the most complicated and difficult years during the time while our Institute was coming into existence and entering the world. It was then that striking ideas and inventions and achievements were "taking off," yet were mixed with incredible discrepancies of our own intentions and determinations compared with the actual results, both in the terms of the time spent and the parameters achieved.

As an injector for the VEP-1 device, Andrei Mikhailovich suggested using the B-2 accelerator with its helical storing of electrons and betatron acceleration, then ending with a synchrotron acceleration stage (shortly before, this latter accelerator, constructed on the basis of Andrei Mikhailovich's ideas, had obtained record currents of circulating electrons, approaching 100 amperes). In order to attain the required energy, about 50 MeV, it was necessary to place in the chamber a sectioned resonator, fed from a pulsed high-frequency generator, and to make a complementary power supply for the further increase of the magnetic field. The first good results of synchrotron acceleration to several MeV energy were already obtained in the first months of the work (this material became the basis of my Master's Degree thesis, defended at the end of 1958). Shortly thereafter, the first successful one-turn extraction of electrons from the accelerator was also achieved (at the time, this was a difficult problem, which no one had previously tried to solve, since the period for one turn was 10 ns).

But there were also some unpleasant incidents. For example, at the end of 1958, for the upcoming International Conference in 1959, it was decided to obtain an electron beam of full energy in the B-2, and then "transfer" it into an experimental one-track storage ring, to study the long life of the stored beam. Doing this would be a world-first in the field of accelerated beam storage rings, and successful operation of the storage rings was a key step in the solution of the colliding beam problem. It was supposed to drive the circulating currents in the storage ring to 100 amperes. We had to work all this out, since the VEP-1 storage ring itself (which was already a two-track device) had been urgently constructed and produced at the Turbogenerator

Factory in Novosibirsk, in order to be then moved to Moscow for the experiments.

Not only was this work strenuous, but most of us who participated lived for practically nothing else. We managed to get iron and coils from the "old" cyclotron and set them up in the same small hall near B-2 in a former medical clinic; we made a vacuum chamber for the storage ring with internal shaping magnets; we developed and made an electron optical channel from the accelerator to the storage ring; and we developed all the other systems for the accelerator and the whole complex. However, not only were we unable to obtain any of the projected results for the 1959 Conference, but formally we went a step backwards from the achievements that had been obtained a year before — practically nothing worked the way it had to. As a consequence, having suffered for almost three years with this facility based on an experimental storage ring and having rebuilt many of the components and systems over and over, we had to terminate it: VEP-1 came from Novosibirsk to Moscow, and all our efforts were now concentrated on starting up (once again with the utmost urgency and again with an eye to the next conference) the whole facility with colliding beams. Again everyone applied superhuman efforts, and, as a result, in the second half of 1962, the facility, which had not yet even begun to "breathe," was moved into the shielded Block No. 3 of the main building of the Institute in Novosibirsk; the construction of the block was not yet completed at that time. Already in the summer of 1963, we obtained in the storage ring the desired long-lived beam that was so necessary for us. The report concerning this work, which was given at the International Conference on Accelerators in Dubna in parallel with the Stanford-Princeton group's report, became for many of us, including me, our first real scientific publication; it had taken us almost six years of the most intensive work! Many persons who had begun the work with us became discouraged and left, especially since being involved in the work entailed moving from Moscow, from the famous Institute of Atomic Energy, to a nonexistent, absolutely "nonguaranteed" institute in Novosibirsk, and also because the real results so terribly contradicted our plans and determinations, as I mentioned earlier. I consider my own standing firm in spite of doubts and temptations during these years as one of my major moral achievements. It was only in 1963 that I began to feel that we had delved to the very depths and that we could actually manage the problems we had posed.

Naturally, in the preceding when I was describing "our" plans, solutions, and so on, I meant Andrei Mikhailovich first of all. He was the center of our whole life. His energy, optimism, drive, inventiveness, and wonderful "physics-ness" at that time comprised the main stimulus to continue the work in spite of all the shocks and disappointments. Soon it turned out that we had already learned much in work and in life and had done it at a reasonably fast pace. Bringing equipment that never worked and that needed the most radical improvement into an empty, unfinished experimental hall, we managed in 1965 — in less than three years — to get the first experimental results on electron-electron scattering. In addition, we developed, constructed, and "brought to life" a novel accelerator facility that was much more complicated than any other previously existing one anywhere else. Also, the experimental results on the physics of elementary particles were obtained on it at the same time as at Stanford.

During our trip to the U.S.A. and to France the following year (1966), Andrei Mikhailovich liked to say that, unlike the Stanford-Princeton experiment, our work had been done by a rather "green" group of people, who, on top of everything else, had had to move from Moscow to Novosibirsk, whereas the American physicists had not needed to move from California to Alaska. Whenever he said this, his audience always laughed and applauded in a very friendly and good-natured way.

In parallel with the work on VEP-1, in 1959 development of the VEPP-2 facility with electron-positron colliding beams was begun. Only later did we learn that work on the same subject had also been started in other research centers. If the creation of electron-electron colliding beams had seemed quite dubious to most experts and if, in any case, it seemed unattainable for our group which was just taking shape, then our discussions about electron-positron experiments were taken by most people as proof of the absolute foolishness of Andrei Mikhailovich and us all. When Andrei Mikhailovich gave a short note about this project to I.V. Kurchatov, the latter sent it to be reviewed by three persons who were considered to be the leading experts in those days.

Paris, Montmartre: the only photograph of Andrei Budker taken abroad (1966).

All three became very involved and interested in their reviews, because the potential opportunities of electron-positron experiments already appeared to be very significant in those days, but all of them gave very negative reviews, saying that this proposal is totally unrealistic and the people who discuss it are groundless dreamers.

Nevertheless, I.V. Kurchatov supported the Institute and pushed through the authorization needed to build the VEPP-2 facility. In 1967, only two years after the first electron-electron experiments in Stanford and Novosibirsk, our group, which grew stronger and larger in leaps and bounds, was able to perform on VEPP-2 an experiment — the first in the world! — on electron-positron annihilation into pions in the ρ-meson resonance energy range. By that time, though, the Lenin Prize — the highest in the country — had already been awarded to our Institute for developing the method of colliding beams for experiments in elementary particle physics. Incidentally, one (but only one!) of the people who had given negative reviews, academician V.I. Veksler, when he visited the Institute after our first results, publicly and fully admitted to having been wrong and congratulated us on our success.

Simultaneously with the ongoing progress in developing electron-electron and electron-positron colliding beams, Andrei Mikhailovich was vigorously searching for practical ways to implement proton colliding beams. At the time, the record energy that had been achieved for protons was in the range of 10 GeV. Andrei Mikhailovich wanted to design a compact and comparatively inexpensive pulsed proton accelerator with a magnetic field one order of magnitude larger than that for usual iron-core magnets and to perform experiments with colliding proton beams at an energy of 2×10 GeV (this energy would significantly exceed that to be available in the accelerator then being designed for Serpukhov[2]). This idea was never implemented, but on the other hand, during this work, Andrei Mikhailovich made two interesting and important suggestions. In order to obtain sufficiently high productivity (or "luminosity," as it was later called) of the pulsed proton-proton machine, it was necessary to learn how to make high circulating proton currents and how to compress these beams down to very small cross-sections. For storing large proton currents, Andrei Mikhailovich suggested the use of charge-exchange injection (namely, first produce negative hydrogen ions and accelerate them in the injector, then bring them appropriately into the injection orbit of the accelerator and "strip" them on the gas target — in this scenario, the physics allows injections to occur during a thousand turns, with a corresponding gain in the circulating current. This "charge-exchange" program was carried out at the Institute with great success, and now all of the largest proton accelerator facilities around the world have gone over to this method, which permits a considerable increase in the intensity and an improvement in the quality of the accelerated beams.

The second step was even more revolutionary. For compressing the proton beams after the injection and the first stage of acceleration, Andrei Mikhailovich put forward the idea of electron cooling. As soon as it had become clear that reducing the size and the energy spread

[2]Trans. note: Serpukhov is a town on the outskirts of Moscow, where the Institute of High-Energy Physics is located.

of a proton beam by the use of an intense "cold" electron beam did not contradict the basic laws of nature, it was suggested at the Institute, in 1965, to use electron cooling for storing antiprotons and creating proton-antiproton colliding beams. During the trip in 1966, which I have already mentioned previously, this manifold of plans and ideas was publicly described to the international physics community. It caused very great interest. Then, at the International Conference in 1971, the Institute presented a design for a proton-antiproton facility, with all the physics and technology aspects described in sufficient detail. Furthermore, another method, so-called "stochastic" cooling, was shortly thereafter invented at CERN. Despite all this, no one except us took up the development of proton-antiproton projects until 1974, when electron cooling was experimentally demonstrated at the Institute. From that moment occurred an "explosion" of interest in this direction, and in the years that followed, the largest proton accelerators at CERN and Fermilab were transformed into proton-antiproton facilities, which produced a whole stream of extremely interesting information about the physics of elementary particles. I very much hope that the proton-antiproton program will at last be accepted as a part, an important priority part, of the very large accelerator-storage ring facility to be constructed in Serpukhov.

At the end of 1969, Andrei Mikhailovich became seriously ill (due to a heart attack). He had had problems with his heart for a long time, but his temperament did not let him pay attention to his health. In fact, Andrei Mikhailovich was removed from the ongoing life of the Institute for about one year, but as soon as he felt a little better, he immediately threw himself back into the current affairs of the Institute and its future prospects in a very energetic manner, practically the same as his earlier extremely strenuous and initiative-taking style. Even though he never felt very well and often went to the hospital for medicine, he busied himself with the greatest enthusiasm, especially in long-range scientific and organizational questions, continually saying that one should always plan how to live, not how to die.

Here, perhaps, it would be of interest to make some comments. In spite of my great esteem (and even veneration) for and my very good relationship with Andrei Mikhailovich in general, some features of his behavior and mannerisms made a hard or unpleasant impression on me. However, in the years after his illness, almost all of these things disappeared, and his wisdom and vitality became his most noticeable features. However, it is not impossible that the effect pointed out by Mark Twain contributed to this change in my perception: the author said that when he, Twain, was 20 years old, his father had very limited intelligence, whereas 10 years later, his father had become considerably wiser, with this process continuing further with the years.

One of the administrative measures for maintaining morale, which at Andrei Mikhailovich's initiative was taken at the Institute at the beginning of this period, was a considerable expansion of the Round Table system. From 1963, all of us, the leading members of the Institute at that time, the members of the Scientific Council, gathered everyday at 12 o'clock noon at the Round Table and discussed all kinds of questions about our own and other fields of science, about what was happening in the Institute, in Akademgorodok, in the country, the world, and the universe. Here scientific and administrative ideas took shape, here we discussed both present and future issues of our life, including seemingly quite minor logistical matters. This system is what kept us — and, I hope, still now keeps us — from becoming stagnant and bureaucratic. The situation, however, changed considerably in 1971: In addition to the 30 or 35 persons who were members of the Scientific Council, there grew up new leading scientists who really became of vital importance for the Institute. Moreover, some of the members of the Council had already lost their active roles for various reasons. Therefore, after long discussions, it was decided that the Institute's Council would meet once a week (every Wednesday) and that on the other days, the topical sections within the Council would separately meet at the Round Table. In this way, we tried to involve all the truly leading members of the Institute, regardless of their age (including even some very young people, who had just graduated from the university). This policy enlarged the circle of those who could exert a direct influence on the life of the Institute to nearly 100 persons and, so to speak, maintained our youth.Ten years later, the course of events made us go even farther in this direction, and now about 250 researchers and project engineers are included in the Round Table system.

Side-by-side with the Teacher: on the right, academician Alexander Skrinsky (1973).

Even though this kind of system seems bulky and its procedure of implementing serious matters through discussions at each of the numerous sections (each according to its own speciality) difficult, we consider it to be most effective in practice, a very important and determinative element in our existence.

I think that this is probably not yet the appropriate place nor time to reflect on complicated and difficult moments, both "external" and "internal," in Andrei Mikhailovich's life, especially those in his last years. Besides, many of the problems and troubles that Andrei Mikhailovich and the Institute encountered during 1970-1977 were not directly related to his personal nature and peculiarities.

The last years of Andrei Mikhailovich's efforts in the field of high-energy physics were focused on the search for methods to create facilities for colliding beams with super-high energies. Many approaches and concrete solutions were thought out and discussed. Many of them were rejected as unrealistic or too ineffective, but even these contributed to our progress in this mainstream direction.

"You may very well be right..." (1974).

The main result of the efforts of many years was the method of electron-positron linear colliding beams (the VLEPP project). Already in the middle 1960's, the possibilities of obtaining colliding beams in linear accelerators had already been looked into and evaluated at the Institute (it is precisely this scheme that has now been implemented at Stanford), and ways to build actual linear colliders were discussed in the report of the Institute that was presented in 1971 at the International Seminar on the Prospects for High-Energy Physics held in Switzerland. The VLEPP project, one of whose authors had been Andrei Mikhailovich, was presented by our Institute, shortly after his death, at the 1978 International Seminar in Novosibirsk dedicated to commemorating the sixtieth anniversary of Andrei Mikhailovich's birth. This will be one of the main projects at our Institute in the field of elementary particle physics for the foreseeable future.

Here I must stop. Many questions remain untouched — questions of science, of everyday life, of everything; and among them the very important question of the relationship and unity of fundamental research and applied work, and the activity of the Institute for the national economy. I sincerely hope that I will be able to deal with them in a similar collection of articles on the occasion of the next jubilee of Andrei Mikhailovich.

The Organizer of the Institute

A.P. Onuchin

In the winter of the 1957-58 school year, at the physics department of Moscow University, there appeared an announcement that Budker would give a lecture on the relativistic stabilized beam. At the time I was in my fifth year of studies. From friends who had done their training under Budker's guidance, I heard that he was organizing a nuclear physics institute in Siberia, which would be involved in doing experiments on colliding beams. Experiments on colliding beams would be something fantastic! I awaited Budker's lecture impatiently, wanting to form my own idea about him. Who is this director of the new institute? Would it make sense to throw in my lot with him?

The big physics lecture hall where his lecture was held was crowded beyond capacity. Budker entered, youthful and energetic. He began to speak and the audience quieted down. He immediately riveted our attention. His description of the relativistic stabilized beam was logical, extremely clear, and very animated. The conclusion of his lecture was devoted to talking about the organization of the new institute in Siberia and about its scientific problems and the opportunities for physicists to work there. There were lots of questions, which Budker answered with enthusiasm and interest.

Budker's lecture and personality won me over, and there arose in me the desire to work in his institute. Siberia did not frighten me: I had grown up there, and I liked the people. The opportunity to become one of the participants in the creation of a new institute looked highly attractive. Prior to this, however, there had to be an interview. Would I pass it? Fortunately, everything went all right, and I was accepted into the Institute of Nuclear Physics. At the same time and from the same class as mine, several other persons were accepted by Budker who also still work at the Institute — academician A.N. Skrinsky, professor I.N. Meshkov, professor S.G. Popov, and V.E. Pal'chikov, candidate of physics and mathematical sciences.

It was only many years later that I realized what titanic labors Andrei Mikailovich was putting into organizing the Institute. His giving this lecture at the physics department of Moscow State University was but one of the many examples of his activities. Now, looking back, it becomes clear to me that the creation of the Institute of Nuclear Physics was Andrei Mikhailovich's primary occupation, the work to which he devoted his entire life. Not only did he propose a theory for how to organize a scientific institute, but in fact he created a first-rate institute that is highly regarded by scientists all over the world.

Naturally, Andrei Mikhailovich paid primary attention to issues related to selecting and educating physicists for the new institute. With everyone who wished to enter the work force of the Institute, he had a rather detailed interview, and he was really happy when he was successful in getting a new person. I remember the year 1961, when I moved from Moscow to Novosibirsk. There was a somewhat complicated situation at the Institute, in particular having to do with questions about doing experiments on colliding beams. The VEP-1 and VEPP-2 devices were under construction. B.G. Yerozolimsky, who was the head of the laboratory in Moscow and who was involved with preparing experiments on VEP-1, decided not to move to Novosibirsk. Andrei Mikhailovich persuaded B.V. Chirikov, who at that time was already living in Novosibirsk, to take over these problems. I recall how we walked around with Boris Valerianovich in the shielded experimental hall that was being built and tried to get a rough idea of where to situate the VEPP-2 storage ring and its control room, together beginning to make estimates of the background conditions at the VEPP-2 storage ring. However, Boris Valerianovich, in all honesty and sincerity, said that he did not want to work with such large devices, with all their technical and personnel problems. Then, one day, during his next trip out from Moscow, Andrei Mikhailovich happily told me: "Good news. I found the leader for your laboratory. He worked under Niels Bohr. Now he is working in LIPAN, and opinions about him are not bad at all." This was Veniamin Alexandrovich Sidorov, who at that time had not yet defended his candidate's thesis. I also remember Andrei Mikhailovich's genuine joy when he managed to persuade V.M. Galitzky, S.T. Belyaev, and R.Z. Sagdeev to move to Siberia.

Andrei Mikhailovich spent much time and effort on the preparation of physicists at Novosibirsk University and also involved himself with questions about the admission of students. I remember his farewell wishes to us young physicists when we departed to go to the towns and district centers of Siberia and the Far East Region to run the secondary-school Olympiad. This was the second round of the Olympiad. The first round of testing had been administered by mail. We had to select the best youths for the third round. Andrei Mikhailovich told us: "You probably will encounter the fact that the children have been poorly trained. Try to distinguish talent from training, and on no account belittle the authority of the teachers. Remember, a teacher does a great work — he enlightens, he increases the level of knowledge."

I also recall, prior to the university entrance examinations, his instructions to the members of the physics admissions committee. Again, he primarily emphasized the need to distinguish capability from education. The levels of preparation in various schools and families are quite different. We had to select those with a talent for physics. "Try not to let a young man's appearance, his behavior, and his way of speaking influence you one way or the other. Try to pose problems that allow you to check their physics thinking, not pretentious problems in which they can display their ability to solve brain teasers."

In Andrei Mikhailovich's opinion, the Institute had to be organized in such a way that physicists be the main people in it. The entire structure of the Institute had to be subordinated to this principle. Budker's Round Table — the Scientific Council of the Institute — is well-known. Practically all of the important questions about the life of the Institute were resolved here — plans for scientific projects, the distribution of apartments, the hiring of staff, the status of work on the experimental devices, new ideas, the functioning of the service departments, awards and reprimands, personnel issues, budgetary matters, experimental results, and so forth. The Scientific Council included the leading physicists of the Institute and the heads of the main service departments. When the Institute was small, the Round Table met everyday at 12 o'clock noon. After the number of leading scientists had increased, topical sections of the Scientific Council were set up.

Andrei Mikhailovich perceived the power of the Round Table to be based on its providing the conditions under which the most difficult of questions could be discussed openly at the Institute. The director still makes decisions. But without a Round Table, he would receive his information from conversations with individuals or from members of the administrative staff. In this case, the information is strongly distorted, by either the activity or the passivity of the person. "Activistic" employees are the most dangerous because not only will they come to the director on their own volition, but they will also ask other employees to "happen" to drop by, in order to influence the director's opinion.

Andrei Mikhailovich waged a continual struggle to keep the service departments of the Institute at a minimum. They must do only those things that are absolutely necessary for physicists. If the personnel department, the accounting department, the filing department, and the staff of the scientific secretariat become too large, they will set up such a bureaucratic system of records, controls, office hours, and so on, that the physicists will no longer feel themselves to be the primary people in their own institute, but will be wasting lots of time. So far, in our Institute, the budget planning department has a staff of only one person, and to get an entry pass written for a visitor, all it takes is a call from the head of a laboratory.

Toward the pilot production division and the machine shops, however, Andrei Mikhailovich had quite a different attitude. He thought that the pilot production facilities of the Institute should have powerful capabilities. When talented physicists are brought together, the number of good ideas is practically unlimited. The main limitation on the work is then the capability for pilot production. From the very beginning of the organization of the Institute until the last days of his life, Andrei Mikhailovich involved himself a great deal with pilot production. Somehow he persuaded the director of a large Novosibirsk factory, Alexander Abramovich Nezhevenko, to come to our Institute and work as deputy director. It must be said that he was really a unique find. A.A. Nezhevenko organized a strong production division at the Institute. But what was most amazing was that Alexander Abramovich understood so well what was special about production in a scientific institute. For example, during the construction of a device, the physicists and the director are obviously not satisfied because it is taking too long and they want to have it more quickly. Alexander Abramovich exerts tremendous efforts to accelerate the progress. Then, suddenly, the Scientific Council and the director decide to stop this project, because another solution, more interesting and promising, is found. Alexander Abramovich halts construction on the old device and energetically applies himself to the new one — truly energetically, without being offended, without even asking a "silly" question like "Have you given it enough thought?"

Alexander Abramovich had a wealth of experience in life, and physicists felt warm and comfortable with him. Sometimes we called him "Mommy." Not only was it possible to discuss all kinds of questions with him, but also to obtain real help.

Such was Alexander Abramovich's attitude toward the Institute — and, of course, it was the result of Andrei Mikhailovich's great and tactful efforts.

I wish to relate some of the strong impressions that I had during the first period of my work at the Institute in Siberia. I moved to Novosibirsk with my wife in 1961. At the railway station, we were met by the head of the personnel department, Ivan Anufrievich Yadrov. We were very pleased with such a reception — at the time I was quite young, having only graduated from the university two years before.

The main building of the Institute was still under construction, so the physicists worked in the building for the machine shop, and there were many problems. In this difficult situation, Andrei Mikhailovich established perfect order for how the supply department should associate with the physicists. Every morning, an employee of the supply department came to us and asked what we needed. The next day, she either brought us everything that was needed, or called and asked us to come and get something if what we had ordered was too heavy. If what we needed was not in the supply room, she ordered it and then kept an eye on the progress of the order.

And how important for the physicists was the organization of the express-order group at the machine shop! From simple drawings, this group would complete a 20-man-hour work order in no more than two days. All that was needed was the signature of a scientific group leader.

I also remember Andrei Mikhailovich's struggle against "alcohol." It is no secret that at many institutes, the scientists gave alcohol to the workers to bribe them to complete a job request. Some places are in such a fix that it is absolutely impossible to get anything done without alcohol. Andrei Mikhailovich spent a lot of time preaching preventive "sermons" on this topic, and once at the Institute there was a most unusual directive. Alcohol was not to be kept in supply in the laboratories, but would always be available from the safety department.[1] If alcohol was needed for work purposes, someone from the safety department brought a can and distributed the necessary amount. This order did not last for very long, but psychologically it left a trace for many years.

Now I would like to dwell upon the particulars of Budker's unique style for working on scientific problems. It is impossible to tell about everything because there were so many original and amazing things.

When selecting scientific subjects, Andrei Mikhailovich was of the opinion that it is necessary to take up those problems that are the most important. Science has an unlimited number of problems, but they differ quite a bit in importance. The solution of some problems will bring about essential changes in the development of science and the national economy, whereas the solution of others will be only somewhat useful to understand phenomena or may even be useless and unmemorable. Andrei Mikhailovich chose the physics of elementary particles and controlled thermonuclear fusion as the main subjects for the Institute. Elementary particle physics is on the forefront of fundamental science having to do with the structure of the world of the microscopic; it leads to the most profound changes in our understanding of the nature of many phenomena. In fusion research, Andrei Mikhailovich endeavored to work on a practical problem — the creation of thermonuclear reactors — which he considered to be a major issue.

---

[1] Trans. note: Like a hazardous material!

It was very typical of Andrei Mikhailovich, in solving scientific problems, to search for unconventional solutions. There is a struggle, a competition, that goes on in science. In science it is of no value to be the second to discover a phenomenon. If we take the standard well-explored path, at best we will only be able to chase after our "competitors." However, if we find a nonstandard solution, then it is possible to make a qualitative jump. In this case, of course, there is the risk that the work will not turn out to be successful. If we do not want to take the risk of unsuccessful solutions, we will not take up those problems that are important but difficult. Andrei Mikhailovich also pointed out another difference between the two approaches. Unconventional solutions generate at the Institute an atmosphere of creativity, which leads to the accumulation of talented and inventive people. The conventional approach, however, collects careful people who work according to the "boss-subordinate" principle.

In choosing unconventional paths, he always was on the lookout for those with promise for solving other problems. He went after those directions that were beyond the capability of many other institutes or universities, requiring massive collaboration and involving many people.

All his life, Andrei Mikhailovich searched for unconventional solutions, and he also strove that this style would be present in everything at the Institute. But the most important thing is to find a solution. He often said that our task was not to explain why a certain problem cannot be solved, but to find a solution.

Many examples of unconventional approaches that were undertaken by the Institute can be cited. Suffice it to mention that the Institute pioneered the creation of colliding electron-electron, positron-electron, and proton-antiproton beams; electron cooling; and linear colliding beams. These accomplishments brought world-wide renown to the Institute. Also, not only did the method of colliding beams become generally accepted, but it now "manufactures" most of the results in elementary particle physics. Incidentally, I should remark that in the Soviet Union today, colliding beams exist only at the Institute of Nuclear Physics of the Siberian Division of the U.S.S.R. Academy of Sciences, even though more than 20 years have passed since the time when the first physics results on colliding beams were obtained.[2] These facilities are very complicated and require the solution of many unconventional problems as well as the joint efforts of many physicists, engineers, technicians, and also experts from numerous support departments.

To seek unconventional solutions for fundamental problems, critical components in experimental devices, technology, and the organization of the Institute and its laboratories — this was Andrei Mikhailovich's style of work. For example, it was at our Institute that the world's first accelerators were built that are hung from the ceiling; normally an accelerator stands on the floor. Hanging it from the ceiling permits the use of a smaller size tunnel, which makes the tunnel cheaper.

I think that accelerator physicists and fusion physicists from our Institute will describe many examples of the unconventional and original solutions found by Andrei Mikhailovich. I do, however, wish to recall something about Andrei Mikhailovich's work with detector physicists, of whom I am one. Andrei Mikhailovich was not a specialist in the detection of particles. Nevertheless, while the VEP-1, VEPP-2, VEPP-3, VEPP-2M, and VEPP-4 projects were being developed, he delved deeply into the major problems related to doing the experiments. I remember his statement to the effect that he considered it optimal, when planning experiments on colliding beams, to arrange to put equal efforts into both the storage ring and the detector. He frequently repeated that "detectorists" must search for nonstandard solutions. I recall his discussion with V.A. Sidorov on this subject. Andrei Mikhailovich said that very few unconventional solutions for detectors had been found at our Institute, far fewer than for accelerators. Sidorov responded that most physicists around the world worked not on accelerators, but on doing experiments and developing methods to detect particles. Therefore, there is a greater opportunity to make progress in accelerators. Budker did not like this explanation and continually endeavored to become more involved with methods for particle detection.

One day (this was in 1971), he invited me to his office and said, "Let's do some work on the question of detecting particles. For a long time I have wanted to consider this, but never had enough time. I have no specific ideas. Let's work together and look for directions that no one else has taken...." We began to work. Everyday at about five o'clock in the afternoon, I went to Andrei Mikhailovich's office, where

[2] Trans. note: This statement, written in 1987, is still true in 1993.

he would greet me with pleasure, ask what I had thought up, and begin to tell me what new ideas and thoughts he himself had had, asking me whether someone was doing it in this way or that. Also, why wasn't anyone doing it this way? Is it because it is difficult and complicated? Or just because no one had had such an idea? Our main task, he would say, is to find an approach that no one has come upon. Occasionally we sat up with him until nine o'clock in the evening. That year he was not feeling very well, and sometimes he did not come to the Institute, but summoned me to his home, and we worked there. This was the first time for me to work so closely with Andrei Mikhailovich, and to associate with him was very interesting for me. His style of searching for new solutions was unusual. He would take a phenomenon, consider it qualitatively and semi-quantitatively, and begin to look for completely unexpected ways to apply it. Numerous options were examined, some quite bizarre. A possible solution was compared with the traditional approaches, only to see that it led to but few new results, if any; then, we went on to another one. Sometimes an idea that had already been thought "dead and buried" would reappear in a different guise. Sometimes he would telephone to some expert and ask a question. At last we hit upon an interesting solution, the photodetector, which in today's terminology can be called a planar photoelectron multiplier with positive feedback. This consists of a semi-transparent photocathode and scintillator, between which is applied a constant voltage. Photoelectrons hit the scintillator, producing photons that knock new electrons out of the photocathode, and so on. Positive feedback occurs. This kind of instrument is highly attractive because it takes up very little space and uses thin material, but gives a large signal. In those days there were many problems with the technology of such instruments. Nowadays, this technology has progressed significantly, and it is not impossible that in the next few years these photon counters will be available commercially.

Then our work stopped, because Andrei Mikhailovich once again became interested in new ideas for accelerators. The Institute was setting in operation the VEPP-3 storage ring and building VEPP-2M and VEPP-4, as the search for ways to advance to higher energies went on. The construction of detectors for these storage rings was begun.

I remember discussing with Andrei Mikhailovich the project for our MD-1 magnetic detector. For that era it was a huge detector —

weight 500 tons, magnetic field volume 10 cubic meters, magnetic field strength 16 kilogauss, 300 proportional chambers each about 2 meters in size, 16,000 electronics channels, half a million small wires 30-100 microns in diameter, large gas Cherenkov counters with ethylene pressurized to 25 atmospheres, and a required power of 3.5 megawatts. In terms of complexity and cost, this was an order of magnitude beyond the previous generation of detectors. Budker assembled a panel of people to review this project and, after a brief presentation, asked: "Tell me, please, what is new and unusual about it?" "There are many new features," replied Sidorov. "The field is oriented perpendicular to the orbit of the beam, whereas in other devices it is parallel. It has a huge number of proportional chambers, which physicists are just beginning to master. It has large Cherenkov counters filled with dangerous gas that can explode; no one else has this. The detector has many innovative features, and there is a risk that it will not work at all." Budker liked it. He asked L.M. Barkov about the possibility of superconducting coils for the detector, who responded that the "warm" version was the right choice. Afterwards, Andrei Mikhailovich advised us once more, as follows: "Keep in mind that you are going into a new and unknown range of energy, where there could be unexpected phenomena. Try to choose a sufficiently powerful and efficient detector. If you choose a good one, the Institute will obtain many interesting and unusual results. The work will last for many years. Don't forget about physicists in other countries; they will soon become involved in these problems. Search for and incorporate new approaches."

More than ten years have passed since then. Many problems came up, difficult ones, but we worked with enthusiasm and pleasure. We were successfully able to build the detector, get it operating, assemble a good group of people, and accomplish a large number of physics experiments. Only in Budker's Institute was it possible to have created such a unique and complex detector.

A Unique Style of Work

E.P. Kruglyakov

My first meeting with Andrei Mikhailovich occurred in February of 1958. Along with seven other self-confident (to say the least) students, all graduates of Moscow Physico-Technical Institute, I was invited to the Institute of Atomic Energy (later named after I.V. Kurchatov) for an interview. Actually, however, it would be more correct to call what transpired a slaughter. We were given a tough examination by A.M. Budker, A.A. Naumov, and two of their colleagues, which began at four o'clock in the afternoon and ended, with we students having been completely routed, at about nine o'clock in the evening. For reasons that have always remained a mystery to me, I alone was accepted into the future Institute of Nuclear Physics of the Siberian Division of the U.S.S.R. Academy of Sciences. Of the seven who were rejected, at least three went on to become Doctors of Science, and two received U.S.S.R. State Prizes. Why have I dwelt on this episode in such detail? Because it helps to understand how meticulously Andrei Mikhailovich handled the organization of the Institute. It may well be possible to say that during these years, no one was hired into the Institute without an interview with the director. Andrei Mikhailovich presumed that it was better to reject a "strong" physicist by mistake, than to allow a "weak" one into the Institute.

In later years I had frequent association with Andrei Mikhailovich, but for some reason, it is the impressions from the fleeting encounters with him during the years when I belonged to the category of a "ju venile" that are preserved in my memory with relief and documentary precision.

In the spring of 1959, Andrei Mikhailovich brought academician M.A. Lavrentyev, who was the president of the Siberian Division of the U.S.S.R. Academy of Sciences, to visit the laboratory of A.M. Stefanovsky, with whom I was working (at that time, we were located at the Institute of Atomic Energy in the building of a former medical clinic). A few days before this visit, our technicians had constructed a large aquarium and brought various kinds of fish from their homes. So then Budker brings the guest in, notices the aquarium, and remarks, "How charming!" However, do not think that our director is fascinated by the fish. Everything is clarified by his next remark. Addressing us young physicists, he declares, "Now, then, you can study the topology of electric fields right here in this aquarium." Budker goes on to explain to M.A. Lavrentyev: "If electrodes are placed in the aquarium and a potential difference is applied between them, the fish will invariably orient themselves across the electric field in order to minimize the voltage they feel." The very same day, we did a check of this idea: we placed electrodes in the aquarium and applied a voltage. Indeed, the fish oriented themselves in unison across the field. Regrettably, however, not all of them did — we overdid it with the voltage a little.

What was remarkable in this incident, as in many others, was the extraordinary unorthodoxy of Andrei Mikhailovich's thinking. This is precisely what allowed him to be a continual generator of ideas, ideas that often seemed almost absolutely crazy. Many of these ideas survived no more than 24 hours, but some of them, the "craziest" ones, lived to become embodied in "iron," with the correctness of their originator experimentally confirmed. It is a matter of taste, of course, but of all the suggestions by Andrei Mikhailovich that became a reality, I most admire the beautiful idea of "electron cooling," which Budker first publicly presented at the 1966 International Conference in Paris. Andrei Mikhailovich later told us that the idea of electron cooling had been conceived in 1960, in the "Centralnaya" Hotel, during one of his trips to Novosibirsk. (In 1961 Budker moved permanently to the new Institute of Nuclear Physics in Novosibirsk.) The fact that six years passed from the birth of the idea to its publication seems to have been an exception in Andrei Mikhailovich's life. The idea was so fantastic that no other laboratory in the world undertook its realization. In 1974 it was experimentally confirmed at the Institute of Nuclear Physics of the Siberian Division of the U.S.S.R. Academy of Sciences by N.S. Dikansky, I.N. Meshkov, V.V. Parkhomchuk (all of whom are now Doctors of Physics and Mathematics), and Academician A.N. Skrinsky, together with their collaborators. Nowadays the method of electron cooling is well-known around the world.

It frequently happened that Budker neglected to greet the members of the Institute. One day, shortly after I had been accepted into the Institute of Nuclear Physics, I witnessed such a scene. Budker comes into the laboratory of A.M. Stefanovsky and says, "Tolya, I need you!" Anatoly Mikhailovich, my first scientific advisor at the Institute — a brilliant physicist, who was also known for his quick tongue — scolds the director in a loud voice that everyone can hear, in jest of course (since he knew perfectly well that Budker's brain was always busy working): "Andrei Mikhailovich, how many times do I have to tell you that you should say hello when you enter!" Budker responds without having to think, "Well then, Tolya, if I don't say hello, I am just emphasizing once again that we never leave one another."

How quickly Andrei Mikhailovich could answer any question amazed me many times. I remember an incident that occurred in 1962 when V.A. Sidorov, now deputy director of the Institute and a corresponding member of the U.S.S.R. Academy of Sciences, was defending his Candidate thesis. At that time, there existed in Akademgorodok the United Scientific Council of Physics and Mathematics, which included leading scientists from different institutes. The head of the Council was M.A. Lavrentyev. The defense was successfully coming to an end, with the questions already exhausted, when A.A. Lyapunov — a "pure" mathematician — asked to speak. He asked a question that, evidently, was not addressed to the author of the thesis: "If physicists call this a Candidate thesis, then what do they mean by a Doctoral thesis?" Budker immediately responded, "The same thing, except that the author of the thesis should be a little older, and the thesis itself, a little worse."

Only once did I see Andrei Mikhailovich puzzled by a question. It happened in Novosibirsk, in the summer of 1964. At that time, research in plasma physics and controlled thermonuclear fusion was in full swing at the Institute, and we urgently felt the need for new plasma diagnostic methods, more refined than the ones we had, in order to make progress. I myself happened to participate in developing laser techniques. When he would bring in visitors and show them the various lasers made at the Institute, Andrei Mikhailovich liked to say, half in jest, half seriously, that these were the first lasers in Asia, Africa, and Australia. As a matter of fact, in those days lasers could be found in fusion laboratories only in Europe and North America, and seldom at that. A little earlier, but almost at the same time as the lasers, we had constructed an optical interferometer. This was a unique piece of instrumentation in plasma physics experiments in those days. Ours was the first in the Soviet Union, and the third such device in the world. Andrei Mikhailovich brought many visitors to our group. Among them were the president of the U.S.S.R. Academy of Sciences, M. V. Keldysh; the future president, A.P. Alexandrov; a Nobel laureate, N.G. Basov (one of the three "fathers" of lasers); and many other famous scientists. Among the visitors were also some dilettantes: writers, actors, journalists....

Regardless of who it was that came with our director, he always gave the explanations himself. It was most interesting to listen to his explanations for non-professionals in science: "You can't even imagine how sensitive this interferometer is. Despite the fact that it is very heavy.... Edik, how much does it weigh?" — this question was addressed to me. "About two hundred and fifty kilograms, Andrei Mikhailovich." Budker then continued: "Even though it is so heavy, if a fly landed on it, it would feel it." Usually our guests left taking with them a souvenir, a coin with a hole punched in it by a laser pulse. This was a rarity in those days.... Once, for a French writer, we made a hole in one of his coins. The hole was not in a very appropriate place. "You hooligans," exclaimed Budker cheerfully, "admit it — you did that on purpose!" One day, while I was working with the interferometer, Andrei Mikhailovich brought in academician Boris Pavlovich Konstantinov. He was a professional physicist, and therefore, although Budker's narrative had an emotional flavor (which Budker could not do without), the language of the conversation was scientific. When Budker had finished, Konstantinov asked, "Tell me, please, are you involved with holography?" Let me remind the reader that this was in 1964, when holography had only just begun to develop as an area of science. Andrei Mikhailovich did not yet know this term. "What? Holography?" he asked, repeating the question, trying to stall for time.[1] Then, after a pause that only lasted a couple seconds, he answered, "Only at home."

During the first years of the Institute, Budker carried a heavy load on his shoulders. He was relieved of some of the troubles by deputy directors who were exceptionally well selected: Alexei Alexandrovich Naumov for scientific matters, and Alexander Abramovich Nezhevenko for general matters. The Budker-Naumov tandem had emerged long before the Institute was organized. How A.A. Nezhevenko came to our Institute was told by Budker himself. In the course of his business for the Institute, Budker had met the director of one of the largest industrial plants in Novosibirsk, the Turbogenerator Factory. Budker admired his remarkably vivid personality and skill in leading people. This is how Budker later described the events: "I thought how useful to me it would be to have such a deputy. I gathered up my courage and...proposed it to Alexander Abramovich. Nezhevenko thought it over and agreed." What arguments Andrei Mikhailovich used to persuade the director of a big factory to move to our Institute are not exactly known, but what is well-known is that Budker did not suffer from a lack of eloquence. He could persuade anyone about anything. One of the many items that our new deputy director took upon himself was organizing the Institute's machine shop for pilot production. He managed it brilliantly.

In spite of his successful selection of deputies, Budker's temperament often made him delve into "small details." A well-known example is the enthusiasm with which Andrei Mikhailovich "worked" with the project designers for the Institute's construction. Thanks to Budker, the main building of the Institute — one of the main sights in Akademgorodok — became five-storied. According to the original design, the building was to have had four floors for laboratories and a fifth floor for technical use. Andrei Mikhailovich suggested to the project designers an alternative way, readily implementable, to construct the fifth floor. As a result, the Institute gained more space for laboratories. It was the opinion of B.B. Bereslavsky, who headed the construction project for

---

[1] Trans. note: The term "holography," when pronounced in Russian, calls to mind the two words meaning "nude studies."

the main building of the Institute at the end of the 1950's, that Budker's suggestion improved the architectural appearance of the building. In the 1970's, a building now called the "Annex" was added to the brain center (this was the name Budker gave to the central part of the main building). Already during its construction, it went from being three-storied (according to the original design) to five-storied, and then, thanks to the irrepressible imagination of our director, the "Annex" became six-storied. All this was done by "sight reading," right during the construction process! It was hard on the project designers and the builders, but as a result the Institute had twice as much useful floor space in the end.

V.A. Steklenev, who designed Building No. 13, for housing the largest devices of the Institute, the electron-positron colliding beam machines VEPP-3 and VEPP-4, recalled such an episode. The design work on the building was dragging. The engineers in charge of ventilation and electricity wanted the information that was necessary for the ventilation system and the lighting. It meant that other work on the project would have to be halted for a while. Naturally, Andrei Mikhailovich could not stand it. Addressing the design people, he said, "Why should you spend any time on this? Just put down that the transformer power for lighting the building must be so many kilowatts and for ventilation so many kilowatts, that such-and-such types of ventilators are needed and the number of ventilators is so many." The designers objected that his numbers needed to be substantiated. "Do you need something with a signature?" asked Budker. "Whatever you want! Anyway, afterwards I will say that it was palmed upon me...." In this way, half in jest and half seriously, the work of building the Institute went on. After two months of design work, the same numbers that Andrei Mikhailovich had suggested were derived.

Facing the front of the main building of the Institute, one can see situated on the right-hand side a tall object constructed in an unusual shape, which arouses the curiosity of visitors to Akademgorodok. Here is the story behind this object. According to the general plan for the Institute, a tall ventilation tower was to have been put in the back right corner of the property. The builders, who had gotten behind schedule with other work, applied to the project designers with a request to permit them to place the tower closer to the front of the Institute, in order to reduce the length of the underground ventilation tunnel.

At about the same time, the architect, B.A. Zakharov, and the head engineer for the project, B.B. Bereslavsky, visited Budker to coordinate plans for subsequent work. At the end of the meeting, Zakharov and Budker began to "move" the tower on paper. "Make it red, round, and with a knob at the end, too.... No, that's naughty!" reflected Budker. Bereslavsky, who was bored because he had nothing to do, facetiously suggested substituting a rhombic-shaped tower in place of a round one. Although it was a joke, Andrei Mikhailovich snatched up this idea. They considered one other shape, triangular, which had the remarkable feature that "the tower has equal strength in all directions." At last Budker settled on a rhomb with its corners cut off. Later, copies of this creation appeared elsewhere in Akademgorodok, but our tower was the first one....

Andrei Mikhailovich worked until nine or ten o'clock in the evening on week days, and frequently on weekends as well. He had his own unique style of working. During the many years of the Institute's existence he had come to know each scientist, and any of them could be invited to Andrei Mikhailovich's office for discussions. Usually he would gather together several persons. Who was invited would be determined by the topic of the forthcoming discussion. "Let's work a little," Budker would suggest, and after that would begin a discussion of his latest idea.

Typically, during the beginning stages, the arguments would be presented in the most simple manner, with "hand waving." I remember such a case. During one of these "work" sessions, one of those present went up to the blackboard and began to write something quickly. Andrei Mikhailovich exclaimed, "Take this idiot away! I want to understand the essence of the matter, and here he is doing penmanship with tensors!" Of course, this remark was made in jest, but it emphasizes yet once more that Budker preferred to understand the main point of the phenomena under consideration initially at a qualitative level. What may be said about such discussions? For us young scientists (even though in 1965 I was honored with the title "The Youngest of the Oldest Generation"), it was a wonderful schooling. It did not bother us that many of the ideas "perished" after the first two or three hours. When this happened, we also learned physics. After the "funeral" of his latest idea, Andrei Mikhailovich usually ended the meeting with this line: "Too bad. The idea was so beautiful." There were also more successful days, although less frequent. On these days the final line was: "Well then, the old man can still do something!"

The discussions led by Andrei Mikhailovich were mutually useful. As I have already said, we young scientists learned physics; and as for Budker, he got information about scientific news. He read very little, and hence this was how he learned about most news. Once Andrei Mikhailovich told about a dialogue that took place between him and his teacher, academician A.B. Migdal. The latter expressed surprise that Budker never read anything. The rest of the conversation went this way:

Budker: "Tell me, please, Kadya,[2] do I know physics?"

Migdal: "Yes, you do."

Budker: "But if I read nothing and nevertheless know physics, therefore I must be a genius."

Migdal (after some deliberation): "No, Andrei, you just know physics by hearsay."

Looking back, I can see that, in spite of the everyday whirl of very urgent matters, Andrei Mikhailovich through the years kept in mind the problem of the upbringing of skilled physicists, a problem to which he returned many times. It is known that Budker actively participated in organizing Novosibirsk State University. Everything seemed to be going well. A first-class university that had absorbed the experience of the Moscow Physico-Technical Institute was established. What more was needed? Andrei Mikhailovich, however, continued to search. In the fall of 1961, a group of about 15 young laboratory assistants showed up in the Institute of Nuclear Physics. In the evening they attended classes at the university, and during the day they conscientiously soldered and "tightened nuts," but sometimes during the work day, leading scientists at the Institute would give them lectures. In this way there began a pedagogical experiment of Budker, which was carried out on fourth-year students of top universities in Moscow, Leningrad, and Tomsk. After two years in this program, the students would graduate — not without some adventures related to administrative difficulties — by defending theses and receiving physicist degrees at Novosibirsk State University. I cannot judge how successful this venture was: very few members of the group involved in this experiment remained at the

---

[2]Trans. note: Migdal's given name was Arkady.

Institute.[3] Maybe because of this, a few years later Andrei Mikhailovich asked rhetorically, "You know who should be the first to be sent to the Moon? A Moscovite! All we need to do is just get him there, and back he'll always come by himself!"

In the fall of 1961, one more large-scale experiment was begun, about which the whole country learned in half a year. By a miracle, I still have one of the form letters, now yellow with age, on which these words are written:

U.S.S.R. Academy of Sciences
Siberian Division

To Participants in the First Round of the All-Siberian Physics and Mathematics Olympiad,
Dear Comrade:

The text of the letter explained to each participant in the Olympiad who had been allowed into the second round, the conditions of this next round. The letter concluded as follows: "The winners of the second round will be invited for 45 days to the summer school at Akademgorodok in the suburbs of Novosibirsk on the shore of the Ob' Sea." The signature at the end of the letter was:

G.I. Budker
Chairman, Organizing Committee of the Olympiad
Corresponding Member, U.S.S.R. Academy of Sciences
Professor

The summer school opened on July 10, 1962. Two hundred and fifty youths, selected through a difficult competition, came to Akademgorodok from all parts of Siberia and the Far East Region. Andrei Mikhailovich was the heart and soul of the school. Here I am not referring to the many hours he spent "behind the scenes" at the school, discussing with the young physicists and mathematicians who led the classes for the "Phys. mice"[4] *what* and *how* to teach them, as well as

---

[3] Editors' note: There were three of them: R.A. Salimov (Doctor of Technical Science and the head of a laboratory in the Institute of Nuclear Physics), and A.M. Kudryavtsev and B.N. Sukhina (both Candidates in Physics and Mathematics and also Senior Scientists). [Trans. note: A.M. Kudryavtsev is now also the head of a laboratory in the Institute.]

[4] Trans. note: "P.M.Sh." was the abbreviation for this "Physics-Math School." From the way this abbreviation sounds in Russian was derived the nickname "Phys. mice" for the young students of the school.

solving all kinds of problems, including logistical ones. This side of his activity was not visible to the talented youngsters. They saw another side. Twenty-five years have already passed since then, but even now many participants in the first summer school have great warmth when they remember Andrei Mikhailovich, his wonderful lectures, and the free and easy "conversations by the fountain" in the evenings. From that long ago they still recall with delight a discussion on "Can a computer think?" that took place in the easy-going atmosphere on the shore of the Ob' Sea. During this discussion, each of these "Phys. mice" was able to compete in eloquence with Budker (or, at least, so it seemed to them). I especially wish to emphasize the absence of any barrier between the youngsters and Andrei Mikhailovich. When he associated with them, he did not try to appear magisterial or look like a venerable scholar. I remember one such case. There were about 20 minutes before Budker's next lecture. The "Phys. mice" take a break to play some sports. Everybody is running around, some playing volleyball. We are slowly walking with Andrei Mikhailovich near the house where the children live, discussing the curriculum for the following days. Suddenly Budker impetuously jumps forward and, as he is falling, "gets to" the ball that had come flying from the group of players and sends it back to the youngsters with a precise hit. He then lands on his hands, jumps lightly back up on his feet, dusts off his hands, and winks at me, saying, "Well, then, can this old man still do something?" At the time, the "old man" was 44 years old.

Nowadays, much is said and written about heat pollution of the atmosphere. In those days, this problem did not even seem to exist. In 1962, at one of Budker's lectures, I heard for the first time that, even after the problem of controlled thermonuclear fusion is solved, mankind must limit its appetite for producing electricity because otherwise the planet would face the risk of severe changes in climate and be threatened by the Antarctic ice melting and significantly raising the level of the Earth's oceans. In a fascinating way, he discussed global problems with school children. I am unable to judge whether any priority for bringing up this issue belonged to him, but it was certainly the case that his lectures were brilliant and that the "Phys. mice" listened to him with bated breath. Actually, Andrei Mikhailovich himself also derived great satisfaction from giving lectures to youngsters. Once, following one of these lectures, he formulated in a rather unusual way his attitude toward associating with children: "When before me I see hundreds of intelligent, inquisitive eyes, my feelings are similar to those of a young mother breast-feeding her child...."

In the fall of 1962, after the summer school had ended, a physics-mathematics boarding school was set up in Akademgorodok. This school has continued to this day, with its student ranks replenished every year by the winners of the annual Olympiads. The first of the "Phys. mice" have already long ago graduated from universities and now work at many institutes of the Siberian Division of the U.S.S.R. Academy of Sciences and elsewhere. Some of them work at the Institute of Nuclear Physics, also. Budker's pedagogical experiment was successful, but still he continued to search. I recall that in the spring of 1967, Andrei Mikhailovich suggested to me that I present a course of lectures on general physics. This in itself is a rather difficult course for a beginning lecturer. In addition, however, how about a course that Andrei Mikhailovich proposed to be taught in one year, whereas the physics departments at the leading universities in our country usually devoted two or two-and-a-half years for teaching such a course? Why was it at all necessary to be in such a hurry? Budker's answer to the latter question amounted to this: "Already long ago I noticed that experimentalists mostly operate with equations from the general physics course. How come? Weren't they taught the theoretical physics courses? Of course they were! Then, what's the matter?" Standing up from his desk and pacing back and forth in his office, Budker continued, "The general physics course is privileged in that it is taught during the first year or so, when the students are concerned only about one thing: learning. After two or three years, their brains are occupied with girls. Therefore, we must take maximum advantage of the first years."

The problem of controlled thermonuclear fusion was one of the scientific areas with which Andrei Mikhailovich was occupied and in which he left a shining trail. His first work in this field dates from 1954, when he proposed a method for confining a high-temperature plasma by means of a magnetic trap.[5] Budker's calculations concerning the behavior of particles in the trap were experimentally confirmed by S.N.

[5] Editors' note: Only several years after this, when the cover of secrecy fell off from fusion research, it became known that an American physicist, Richard Post, had come up with a similar proposal in the same year. This plasma confinement system later became called the Budker-Post trap.

Rodionov, a member of the Institute. This became a classic experiment, included in plasma physics textbooks. Because of the fundamental nature of his idea, Budker was awarded a Certificate of Discovery. Incidentally, Andrei Mikhailovich is perhaps the only Soviet scientist who has been twice honored with such certificates (a second Certificate of Discovery was awarded to Budker for his work on the stabilized electron beam).

In 1971, together with the young theorists V.V. Mirnov and D.D. Ryutov, he proposed a novel type of plasma trap, which was named the multiple mirror trap. In this case also, by a twist of fate, a group of American physicists led by A. Lichtenberg independently and practically at the same time came out with the same idea, although the proposal of the Soviet physicists looked more preferable: they formulated the equations describing the behavior of the particles in the trap and obtained some analytic solutions, whereas the Americans only presented the results of computational calculations. I myself happened to do experiments to check the theory. The experimental results showed that the theory was beyond reproach in its description of the processes within the trap!

Andrei Mikhailovich was one of the people who stood at the beginnings of peaceful "fusion." I had an opportunity to hear him tell how this work began. Investigation into the problem of controlled thermonuclear fusion took off immediately after the solution of the atomic problem. Those participating in the work were absolutely sure that it would be possible to conquer "fusion" in two or three years. It was only after many years later that academician L.A. Artsimovitch stated that physicists had never before encountered a problem comparable in difficulty to that of controlled thermonuclear fusion. I repeat: When we set out on the road, it was thought to be a matter of two or three years. Only with regard to the method for solving the problem was there any disagreement. Some people advocating using slow processes, whereas others saw that the use of fast, pulsed processes was the shortest path to success. As Budker remarked figuratively, "On this stage there was a struggle between the sharp-enders and the round-enders."[6] A few years passed. It became clear that the solution of the problem was still very far off. Andrei Mikhailovich was one of those who recommended

---

[6]Trans. note: "Sharp-enders" are people who eat a hard-boiled egg beginning from the pointed end. "Round-enders" start eating from the rounded end of the egg.

stopping and instead beginning systematic research into the physics of plasmas. This "fourth state of matter" turned out to be complicated and treacherous, with its behavior requiring long and careful study.

Thus arose the field of science called high-temperature plasma physics. More years passed. In his speech at the close of the International Conference on Plasma Physics and Controlled Thermonuclear Fusion in 1968, Budker remarked that, although the study of the physics of plasmas was far from being finished, the experience that had been accumulated by the international community of physicists was enough to begin at last the solution of the principal problem — the creation of a thermonuclear fusion reactor. This speech of Budker made a division between two eras: the preceding era of "pure" plasma physics, and the era of goal-oriented studies of reactor plasma physics, which began thereafter.

Many times and by many different audiences, Andrei Mikhailovich was asked when the problem of controlled thermonuclear fusion would be solved. His answer went like this: "If it is of extreme urgency, the problem will be solved the same way the atomic problem was solved, and a reactor will be built in 10 to 20 years. Otherwise, it may drag on for a very long time."

A certain Soviet astronomer, famous for discovering volcanoes on the Moon, published a small number of copies of a very far-fetched book on mechanics. Suffice it to recall, for example, that in his description of mechanics, time could convert into energy. Budker was once asked his opinion about it. His reply was: "If I were to be completely released from all duties, I could manufacture a dozen such theories, which, moreover, would not be self-contradictory. A new theory is needed only when experimental facts arise that contradict previous theory."

Scientist and Director

A.G. Khabakhpashev

Andrei Mikhailovich Budker was a brilliant, very versatile person, and writing about him as a physicist and as an individual is an interesting and engaging task, although difficult even for a professional author or journalist. It is all the more difficult for someone who does not have such expertise. Therefore, here I will attempt to describe only two aspects of the diverse activities of Andrei Mikhailovich — his views on the principles for organizing and managing a scientific institute, and his talent for being innovative. The latter characterization is very relative and may not be quite apt. Andrei Mikhailovich had an extremely fertile mind. To him belongs the credit for some great scientific discoveries, as well as for numerous original technological solutions, inventions, and successful finds. This was all very typical of the creative method of Andrei Mikhailovich — which will be my point here.

At the end of the 1950's, the design of the first colliding electron beam device was begun at the Institute of Nuclear Physics, which at that time had just been created and was still entirely located in Moscow. For decades, both large and small accelerators had always been constructed so that the particle trajectories were in a horizontal plane. This seemed natural. Andrei Mikhailovich suggested standing the VEP-1 device[1] on end and then proved the utility of this unusual approach. The vertical orientation of the device did not provide any special benefits to the physicists directly involved in constructing the storage ring. However, for doing experiments that investigated the angular distribution of scattered electrons, the vertical orientation of the orbit plane had indisputably significant advantages. In particular, this was precisely the way that the background cosmic radiation was considerably reduced.

Many years later, while VEPP-4,[2] the largest electron-positron storage ring at the Institute, was being constructed, Andrei Mikhailovich suggested that the magnets be hung from the ceiling of the tunnel, which was 300 meters in circumference, rather than be placed on the floor, as had been done when all the other accelerators were built. This was highly unusual, but with this scheme a lot of space was freed up in the tunnel, so that, despite its modest size, electric cars went through the tunnel easily, cables were comfortably run through it, and so forth. Of course, these two examples are far from being the most important and significant instances of the creativity of Andrei Mikhailovich Budker, but they are sufficiently typical of his general approach to the solution of each problem he took up. Morning, afternoon, and evening, in his office and in meetings of the Council, Andrei Mikhailovich was always inventing things, always thinking up ideas, always looking for new ways to accomplish each task. Using a traditional approach did not satisfy him, and moreover it would cause something like protest from him.

In this respect, Andrei Mikhailovich's view on how a scientist should take up new work is very interesting. Never begin by studying the literature! First, mull over the problem by yourself; come up with one or more of your own solutions; weigh their strengths and weaknesses. Only then should you look up this question in the literature, compare the approaches described therein with your own, and then choose the best one. If everything began from the literature, nothing new would ever be done; at best, old ideas would merely be modernized.

---

[1] Trans. note: VEP is the Russian acronym for electron-electron collider.

[2] Trans. note: VEPP is the Russian acronym for electron-positron collider.

To all of this, one more secret of Andrei Mikhailovich's "cooking up" of inventions must be added. Whenever he suggested a new solution, an estimate of its effects was immediately calculated. Here, Andrei Mikhailovich always assumed parameter values that were optimistic — that is to say, in his favor. This always caused protests from our side, the members of the Council. But we were wrong. Andrei Mikhailovich had a very simple explanation for his optimistic overestimates. If we calculate the value of the effect conservatively, strictly on the basis of what can be guaranteed, then, as a rule, a new idea must be rejected since it will not give any gain. Besides which, it is most likely that a useful idea will also be rejected in this way, whereas an overestimated evaluation will show that the idea is feasible and that it must be worked on, only after which can the necessary gain be obtained.

Time and again Andrei Mikhailovich asserted that we could not compete with the best high-energy physics laboratories in other countries if we followed the standard path. The foreign laboratories have access to all sorts of factory-produced equipment, including powerful high-frequency and pulsed power supplies, complex high-vacuum and electronic equipment, and many other things. Dozens of specialized high-technology companies will gladly undertake tasks proposed to them and will complete them rapidly, with high quality. The customer just needs to have money! Our Institute has to make all of its equipment by itself, except for a limited number of standard instruments. Naturally, in this sort of competition, we will lose, both in terms of time expended and in reliability. Andrei Mikhailovich said that there was, consequently, only one way for us to avoid being behind and even at times to get ahead — which is to look for new solutions of our own, to find approaches that other laboratories do not yet know. Even if new solutions contain a certain degree of risk, they are still preferable to the standard way, because taking the latter surely condemns us to be behind.

This approach is undoubtedly correct, but its implementation requires that conditions be created in which new ideas and projects and technical solutions have an opportunity to be realized in practice. Such opportunity could be provided only by our own Institute, the Institute envisioned by Andrei Mikhailovich. Consequently, it also led to the natural desire to move to Siberia, where the new Siberian Division of the Academy of Sciences was being formed.

As a result, the Institute of Nuclear Physics was created in Novosibirsk — a perfect place to implement and to check in practice the numerous ideas and inventions of Andrei Mikhailovich. The most significant of these are well-known. First of all, it is necessary to mention here the practical implementation of the method of colliding beams, which required many fundamentally new solutions and the development of new technologies. Andrei Mikhailovich proposed a scheme for extracting the electron beam from an accelerator and trapping it in a storage ring. He also developed and implemented a method for increasing the efficiency of collecting positrons, with the aid of magnetic parabolic lenses of short focal length. Soon thereafter Andrei Mikhailovich put forward the new idea of using magnetic lithium lenses to focus electrons on a target and to collect positrons efficiently. (Incidentally, lithium lenses made by INP staff members are now being successfully used for focusing protons and antiprotons at Fermi National Laboratory in the U.S.A. and at CERN in Switzerland.) Andrei Mikhailovich invented a high-frequency voltage generator, the gyrocon, as a power supply for accelerators and storage rings. This generator, which is based on an original principle, is distinctive because of its high efficiency. Several modified versions of such a generator were later built at the Institute.

While still in Moscow, Andrei Mikhailovich had proposed the idea of magnetic mirrors for confining plasma in open axisymmetric traps. This work — and also another, the self-stabilized electron beam — were listed in the official Soviet registry of scientific discoveries. Work on investigating several modified traps with magnetic mirrors began in Novosibirsk. In the 1960's Andrei Mikhailovich suggested the fundamentally new idea of proton cooling, in other words, methods to damp the amplitude of the proton transverse oscillations with the use of an electron beam. Experimental verification of the electron cooling method, carried out at the Institute in the 1970's, permitted the question of creating devices with proton-antiproton colliding beams to be seriously posed. This list can be continued further.

Very soon after the Institute was organized, numerous groups of Soviet and foreign scientists began to come to Novosibirsk more and more often. With great interest they became acquainted here with many new works, already carried out or just conceived. Andrei Mikhailovich was certainly ambitious in the good sense of the word. He was very pleased by the sincere admiration expressed by visiting colleagues for the success of the Institute, success that was usually obtained due to novel, original solutions.

For whatever he turned his attention to, Andrei Mikhailovich had a sporting enthusiasm — to think up something on his own and to do it better than it had been done before. Of course, this mostly concerned physics, but not exclusively. Here is an example. Andrei Mikhailovich took pleasure in "inventing" a new ventilation tower: instead of a conventional round tower, which surely would have spoiled the facade of the Institute, he came up with a different and original shape that fit in very well with the appearance of the building. Moreover, he was very proud of having done so.

Andrei Mikhailovich liked to tell others about his new ideas. Usually the Scientific Council was the place for these discussions. But all too often, unfortunately, when he described his new suggestions, we dampened the pleasure that Andrei Mikhailovich felt. For some reason, we more often focused our attention on the weak aspects of the solution, on the difficulties that might be encountered during its implementation, and so forth, than on the originality and beauty of the general idea. Critical analysis is of course necessary when something new is discussed, but it is not impossible that in this case we overdid it to some extent. Nevertheless, even our excessive scepticism did not diminish Andrei Mikhailovich's interest in associating with us. If he met any one of us during his evening walk (which he began to do only after his heart attack, out of necessity), he immediately began to discuss whatever new problem he was concerned with at the moment.

Andrei Mikhailovich's thinking was remarkable and original not just in physics and technology. A great variety of questions were often discussed at the Council, including issues quite removed from the subjects of the Institute. These might be problems of bringing up children, a movie, school education, family relations, relations between men and women, social life, and many other things. Andrei Mikhailovich participated with pleasure in such discussions and often instigated them. Even in cases when our opinions differed with those of Andrei Mikhailovich, I cannot help mentioning that his judgments were interesting and nontrivial and involved his own approach, paying attention to those aspects of an issue without which it was impossible to find the right answer. Such discussions were highly fascinating and beneficial for us.

In distinguishing the main point of Andrei Mikhailovich's views on the organization and management of the Institute, it is probably necessary for us to dwell upon three points — the management of the Institute through the Scientific Council, the constant influx of young scientists into the Institute, and the competent pilot production facilities.

Andrei Mikhailovich said many times that even before the Institute of Nuclear Physics was created, he had for many years labored over the theory of organizing and managing a scientific institute. The director cannot manage an institute by himself; for this purpose, therefore, he needs to set up an administrative staff — deputies in all of the major areas; departments for personnel, planning, supplies, and maintainence; and so forth. The director acquaints himself with the main scientific results at meetings of the Scientific Council and during visits to laboratories; all other information about the life of the institute, about the functioning of its services, even about personal relations within the institute, is obtained from the administration. Also, the actual managing of the institute is done by the director through the administrative staff. However, this management style, as Andrei Mikhailovich pointed out, has two main shortcomings. The director receives his information about the institute's work not from those who are directly occupied with scientific endeavors, but from his assistants, no matter how good they are. Also, all the service departments of the institute plan their work so that top priority is given to meeting the requests of the director and his assistants, which, in general, they succeed in doing. At the same time, very little attention is paid to the needs of the scientific laboratories and groups.

As a counterweight to this style of doing things, Andrei Mikhailovich implemented in practice another method — the management of the Institute through a Scientific Council. He recommended, as a first step, that the Council gather every day at 12 o'clock noon to have a cup of coffee at the Round Table, which was specially constructed for this purpose. Initially this suggestion aroused objections from members of the Council. Meeting every day would interrupt work — this was no good! Andrei Mikhailovich responded by suggesting that the Council try this schedule for one month, to begin with. If, on a given day, it was inconvenient for someone to attend due to work, he may miss the meeting. He may also come late or leave early. Then, after a month, we will decide what to do. But after a month had passed, no one even mentioned the need to cancel having meetings every day. Everyone experienced practical benefit from the Council working this way.

At the Scientific Council — or, as it is more commonly said at the Institute, at the Round Table — all the main issues of the work and the life of the Institute are discussed. These include plans for new scientific work, as well as the most important and interesting results that have been obtained in the Institute's laboratories. Here, at the Council, the topic can also be trips to conferences, visits to other institutes in the U.S.S.R. and abroad, the most recent work of world laboratories in the fields of high-energy physics and controlled thermonuclear fusion, and the most interesting physics results in general.

At the same time, the Council discusses many logistical and technical questions — the work of the machine shop for pilot production, the operation of the supply department and the other service departments of the Institute, anything that causes difficulties in the everyday work of the laboratories. As a rule, these questions are not planned in advance, but are brought up by members of the Council in accordance with the perceived need. Even such an important social issue as the construction of housing is invariably considered at the Council.

With the Council functioning this way, the director has the opportunity to obtain all the information about what is going on at the Institute not just from his assistants, but also from the leaders of laboratories and groups, that is to say, from those persons who directly conduct the research. All orders and instructions are discussed, and often modified, by the Council before they are announced. Questions about personnel, raises in salary, and awards are also considered at the Council, not only in connection with the scientific laboratories, but also for the service and technical departments of the Institute. The opinion of the Council members about the functioning of the services is especially useful to the director in making proper decisions, since above all else the operation of the service departments, be it good or poor, affects the work of the scientific laboratories. Moreover, the constant contacts and association between the laboratory heads and the leading scientists creates a good atmosphere at the Institute, something the director is also interested in having.

Andrei Mikhailovich had no patience to listen to long speeches by others, so he often interrupted those who presented reports. However, he had a wonderful ability to precisely grasp the opinion of Council members concerning the question under discussion and to understand the situation from individual remarks, notes, and reports that he did not hear to the end, which ability enabled him to find a correct decision. It may be said that the Council gave him the opportunity to monitor the pulse of the Institute. If a suggestion of his did not receive support from the Council, Andrei Mikhailovich did not make a decision. In cases when he thought a decision was important, he began to agitate for it, to persuade, to substantiate his point of view. And more often than not he was successful. There were some cases, however, when the Council gave in because it was exhausted by the energetic and lengthy pressure from the director.

The working style of the Council, suggested by Andrei Mikhailovich, was also undoubtedly useful for laboratory heads and group leaders. Time spent in meetings that were held more often than usual paid off entirely. Most questions whose resolution required going to the director or to his deputies and assistants now could be dealt with directly here at the meeting. A laboratory head knows what and how things are being done in other laboratories and in the Institute as a whole. It lets him better plan his work and support a common approach to complicated issues like promotions to higher positions, presentation of theses, and so forth. In speaking about how laboratory heads benefited from participating in the work of the Council, it is impossible for me not to emphasize the following point. Andrei Mikhailovich led the work of the Institute mostly through the Council, and participating in it was, for the Council members, a good training in how to manage a scientific body.

It is difficult to find another institute where the administrative staff and service departments paid as much attention to the requests of the laboratory heads as they did at the Institute of Nuclear Physics. This was so not just because the Scientific Council had the last word on evaluating their work and that of their heads, but also thanks to the fact that there is at the Institute an appropriate atmosphere, in which everything must be subordinated to establishing the best conditions for the work of the scientific departments. Andrei Mikhailovich liked to remind us that the less we hear about a service, the better it is functioning. This approach was actually an important criteria whenever the operation of the service departments was evaluated. When

Andrei Mikhailovich wanted to explain to visitors his principle behind the Institute's management, he would say in jest something like this: "If I surround myself with administrative staff, they will naturally turn their faces to me and turn their backs (to put it delicately) to the Council members. But to the contrary, if I surround myself with members of the Council, then the administration, which is outside the inner circle, will still turn their faces to me, but now they will also have to have their faces turned toward the Council members."

Andrei Mikhailovich had a negative attitude toward increasing adminstrative staff and services, presuming that it would lead to bureaucracy, excessive over-management of the Institute, and, consequently, to a degradation of the creative atmosphere. For the same reason there was only one secretary at the Institute and for a very long time obviously not enough typists, which often caused delays in publishing papers. The statement about the need to limit administration and services, although correct in general, here had a negative effect. The Scientific Council finally managed to make Andrei Mikhailovich change his mind, and a good typing pool was set up at the Institute. Nowadays it is never discussed anymore, except when the question of giving bonuses is considered.

Andrei Mikhailovich frequently commented that, just like living organisms, all institutes and scientific organizations are subject to growing old. He cited examples when a newly founded institute produced brilliant results, but after 20 years when the institute and its employees had grown old, its work became very average. Andrei Mikhailovich believed that it was necessary to take this very objective process into account and to take special measures to weaken its effect. One of these measures is the constant influx of young scientists, along with a system for "flow-through" of scientists at the Institute. There was no problem with the influx — Novosibirsk University, with active participation by Institute staff members, prepared good candidates for scientists. Things did not go as well with "flow-through," although in most cases the Institute did get rid of weak scientific employees in one way or another, and Andrei Mikhailovich gave his full support to this process.

When among the younger scientists there arose a significant number of talented people, Andrei Mikhailovich reorganized the Scientific Council to enlarge the circle of persons who had regular and direct contact with the director and who exerted influence on the life of the

Institute. For this purpose, three sections of the Scientific Council were organized: elementary particle physics, accelerator physics, and controlled thermonuclear fusion. Furthermore, some of the most talented young employees were included in the Scientific Council of the Institute. Each section met once a week at noon at the Round Table, while the Council of the Institute met twice a week. In this way, the circle of staff scientists who participated in meetings of the Council or the sections became more than twice as large. There was also an attempt to energize the activity of the all-institute seminar, but this was not a complete success.

Andrei Mikhailovich assumed that the laboratories in a scientific institute must depend upon who is the leader, not upon what is the subject. In his way of thinking, this meant that if the head of a laboratory left or was removed for some reason, the laboratory must be disbanded. If there is a member of this or another laboratory who could be a good leader, a laboratory should be set up for him earlier, without waiting for a vacancy. In most cases, this principle was followed at the Institute. Andrei Mikhailovich often discussed the question whether it would be expedient to organize temporary scientific collaborations, laboratories, or groups for the solution of a particular problem and then disband them after its completion. This approach allows the selection of a leader for a certain task, based on his personal qualifications at that time, not on the position he already holds. Besides, temporary groups are more easily able to be freed from being tied to the "coat tails" of older work, which often persist for a long time when a laboratory is organized in a traditional way. Andrei Mikhailovich also considered temporary scientific collaborations to be a highly effective measure for preventing the Institute from growing old. However, this idea was not put into practice at the Institute of Nuclear Physics.

Andrei Mikhailovich's opinion that a machine shop for pilot production at the Institute was absolutely necessary for effective scientific research can be well illustrated with several facts. The construction of the Institute of Nuclear Physics in Novosibirsk began in 1958. Naturally, Andrei Mikhailovich was impatient to get into the first complex of buildings as quickly as possible and start working on lots of his ideas. Nevertheless, the first thing he built in Novosibirsk was a big building for the pilot production machine shops ($6,000$ m^2), not a laboratory building. Here is one more typical example. In 1960 Andrei Mikhailovich went to the director of the Novosibirsk Turbogenerator Factory in order to negotiate an agreement for the factory, with its large-scale equipment, to produce the magnets for the electron-positron colliding beam device VEPP-2. They met for their appointment. During their conversation, Andrei Mikhailovich observed how the director of the factory worked, how quickly and clearly he solved questions that arose, how perfectly he understood the engineering and technical production problems. After one or two such appointments, Andrei Mikhailovich got up the cheek, so to speak, to propose to the director of the biggest factory in Novosibirsk, Alexander Abramovich Nezhevenko, that he move to the Institute as his deputy! Andrei Mikhailovich, as always, was vigorous in his persuasion, very attractively describing the brilliant future that the Institute of Nuclear Physics would have, especially if a smoothly operating large machine shop were to be organized. And, Alexander Abramovich agreed. It has been said that the fact that Alexander Abramovich had heart problems then and that doctors had recommended that he change his place of employment played some role. Be that as it may, he came to the Institute of Nuclear Physics and organized a strong machine shop and many other technical services. Andrei Mikhailovich's choice was very successful.

Even during the last years of his life, when a relatively large machine shop was working smoothly (22,000 m^2 floor space, approximately 700 employees, more than 200 machines), Andrei Mikhailovich exerted active efforts to build or to get a mechanical factory in Siberia or even in the European part of the U.S.S.R. The goal was the same: to increase the pilot production capability of the Institute considerably and to have the opportunity to make and deliver for the national economy and for sale abroad accelerators for radiation technology, high-power synchrotron radiation sources, and other equipment, in the development of which the Institute has had much scientific and technical experience. Later, this problem was solved without such an additional factory. The industrial parts of the Institute devoted to pilot production were enlarged to twice their size. Decisions were also made to construct housing, without which it is very difficult to recruit the necessary personnel.

The examples described above emphasize the attention that Andrei Mikhailovich paid to pilot production. He fully understood that all or nearly all of his best ideas and those of his colleagues would become hopelessly obsolete and lose all of their value if the Institute, occupied as it is with problems such as high-energy physics and controlled thermonuclear fusion, were to rely on industry or even on the pilot production factory of the Siberian Division of the U.S.S.R. Academy of Sciences to build its extremely complex and varied devices.

In the middle of the 1970's, the pilot production facilities of the Institute had a strength of one million man-hours annually. In addition, a quick-order machine shop was set up with an annual strength of 100,000 man-hours. At the quick-order shop, orders were filled in one to three days. Considerable attention was paid to how both of these kinds of hours were distributed among the laboratories. Besides all this, each of the laboratories had several drills, lathes, and milling machines. There was always a correlation in the Institute's laboratories — the number of technicians and mechanics was approximately equal to the number of scientists and engineers. All this helped to shorten the time for producing, assemblying, and getting a device going, and allowed engineers and scientists to be freed from work that could be done successfully, and sometimes even better, by technicians and mechanics.

The principles for organizing and managing the Institute, which Andrei Mikhailovich proposed and implemented, are very well-founded. So far they have not been subject to much modification. This is not only our duty to the memory of Andrei Mikhailovich Budker, but also the best confirmation of the great effectiveness of the Institute's organization.

As If It Were Yesterday...

T.A. Vsevolozhskaya

My first acquaintance with Andrei Mikhailovich Budker took place in an informal setting — at one of the institute evening parties at the House of Culture of the I.V. Kurchatov Institute of Atomic Energy, in the fall of 1959. Not long before this I had married a young man who was an employee of the Institute of Nuclear Physics, and, as for myself, I was just about to graduate from the physics department of Moscow University. At the party, Grisha (my husband) introduced me to Andrei Mikhailovich, who, in the manner of a good-natured but concerned father, chided him for getting married without asking his director's advice. "Keep in mind," he said, "that if you decide to have a son, let me know beforehand." Although said in jest, his words seemed to me to encroach upon my rights, and so I said that I would be the one to make that decision. "But I hope that Grisha will also be involved," quickly parried Andrei Mikhailovich, and I realized that one should be very much on one's toes in a conversation with him.

Among work-related incidents, I especially remember a discussion about versions of a proton-antiproton converter for the NAP project.[1] This was in the winter of 1967-68. At that time there already had been developed and implemented on VEPP-2 an efficient system for electron-positron conversion that used parabolic lenses. Following Andrei Mikhailovich, we referred to these in our discussion as *Ha*-lenses because their concave surfaces resembled the letter "X" (which is pronounced "Ha" in Russian). However, this system could not be simply transferred over to the conversion of protons into antiprotons due to the large size (about 10 cm) of the nuclear target. A few months before this conversation with Andrei Mikhailovich, G.I. Sil'vestrov and I had given thought to and discussed several versions of a converter, including one with current. In this latter version, antiprotons born along the entire length of the target are confined near its axis by the magnetic field of the current, which has practically no effect on the motion of the primary protons due to the great difference in energy between the primary and secondary particles. This permits the efficiency of capturing antiprotons in the storage ring to be enhanced, although the parameters of the converter — a magnetic field gradient of 100 Tesla/mm, with a repetition rate of a few seconds — seemed to us to be absolutely fantastic.

During the discussion with Andrei Mikhailovich, at first the conversation was limited to considering possibilities for using a solenoid; however, its efficiency turned out to be small, even if the most daring assumptions about its electrotechnical parameters were made. Nevertheless, it was this daring that made me suggest a current converter. Andrei Mikhailovich immediately grasped the idea of this converter and burned with enthusiasm. After quickly making the necessary estimates of the efficiency and the parameters of such a converter, we began an animated discussion about the possibility of achieving them. Ultimately, the idea, which had seemed so fantastic, acquired features of reality, and working on its achievement turned out to be of interest. Andrei Mikhailovich was now in very good spirits. The whole discussion proceeded merrily, in a free and easy manner, with joking and laughing. When it was brought up that the parabolic lens for collecting particles from the converter must explode during operation (as must

---

[1] Trans. note: NAP is the Russian acronym for the antiproton storage ring at the Institute of Nuclear Physics.

the converter itself), Andrei Mikhailovich cheerfully said, "So what? We'll set up an automatic machine that will stamp out *Ha*-lenses as fast as required. Something like this: *Ha-Ha-Ha*-..., ha, ha, ha," ending in laughter.

During the next few years, laboratory studies that gave positive results were carried out on the feasibility of making a current converter with a radius of 1 mm and a magnetic field on the order of 100 Tesla on the surface. An explosive parabolic lens that could be replaced in a few seconds was also developed. That particular evening, as I was leaving, while Andrei Mikhailovich, A.N. Skrinsky, and G.I. Sil'vestrov were still continuing their work, the director said to me: "Good for you; drop by more often! You create a good atmosphere!..."

Never Follow a Pattern!

I.N. Meshkov

I myself had opportunity to experience personally the system "according to Budker" for selecting and educating personnel. Later, I heard him many times give his thoughts about how a "regular" scientific research institute is organized: viz., a director is appointed, he recruits a head for the personnel office, and the latter "recruits the staff" by means of applications and announcements.

The north physics lecture hall buzzed with impatience: a famous scientist whose talk had been diligently publicized well beforehand was obviously not distinguished for his punctuality, being late also for this meeting with graduating students of the physics department of Moscow State University. At long last, three persons entered through the door that only the lecturers used. A.N. Skrinsky, who had involved me in this "venture," confirmed that the one in the front was Budker. He was a man of medium height and sturdy build. (Later I learned that in his student years, Andrei Mikhailovich had been keen on gymnastics; at the age of 50, he could still easily do handstands and swim the crawl with excellence.) His large head was "decorated" with a respectable bald patch; somehow this detail made a strong impression on me.

Instead of talking, as we had expected, about the Siberian Division of the U.S.S.R. Academy of Sciences and trying to persuade us to move to Novosibirsk to work there, he immediately began to talk about physics. The subject of his talk was the relativistic stabilized electron beam — one of his own ideas, on which he was at that time working hard with about half of the scientists in his laboratory. After his half-hour talk, at the end of which Andrei Mikhailovich made a few comments about Akademgorodok ("a town in the forest...with cottages, although these are not for you; for you, apartments in nice buildings have been prepared"), he suggested that those who wished should take a written examination. The audience dwindled considerably; then the young men who had come along with Andrei Mikhailovich (their names, as I managed to find out later, were E.A. Abramyan and V.I. Volosov; B.V Chirikov and G.F. Filimonov came later, only for the examinations — the last of these I knew through mountain climbing) distributed three problems to each of the persons who remained and gave us two hours to work on them. My test, as I recall, had a problem about scattering in a Coulomb central field. The problem could not be solved without knowing the theorem about conservation of angular momentum in a central field, but this theorem was largely unfamiliar to graduates of the Moscow University physics department in those years (and I was no exception). However, I managed to get out of difficulty and break through in my own way by giving an approximate solution and obtaining a verisimilar answer. Apparently because of this, I was allowed to take the oral examination as one of the ten "finalists." My examiners were satisfied by how I "straight off" solved the problem of colliding beams and handled the energy-momentum four-vector.

Other than myself, only A.P. Onuchin was able to pass the examination at that time. There was nothing we could do about it — that was how we were taught to think in those days. For example, A.N. Skrinsky, who came to do his practical training in Andrei Mikhailovich's laboratory when he was still only in his fourth year, was unable at this examination to give the correct solution for the drift motion of a particle in crossed electric and magnetic fields. Is this funny? No, it's not. I describe it in such detail in order to give an idea about the level of our preparation at the time, which Andrei Mikhailovich and his colleagues immediately tried to improve when they designed the course of study for the physics students at Novosibirsk University.

As the time approached for my being assigned to a job, I was still hesitating whether or not to go to Siberia, and therefore I decided to go talk with some representatives of one of the Moscow "closed" scientific research institutes. Such institutes we called "boxes" because they had no names, only post office box numbers, due to the secret ("closed") topics of their activity, and nuclear physics was one such topic. Incidentally, this institute was situated not very far from my home district of Kuz'minki (rarely is one so fortunate in Moscow as to be able to work near home). These VIP's (which is clearly what they were, judging from their manners and behavior) interrogated me for a rather long time about my social activities at the university, my sports activities, and so forth...without saying one word about physics. This determined my choice once and for all.

The system of giving an examination to those who wanted to enter the Institute of Nuclear Physics was applied without exception to everyone — from laborers, laboratory technicians, and younger scientists, to those with extensive experience and many years of work. This system is still in operation today. Later I had the opportunity to witness how much attention and care Andrei Mikhailovich paid to the "staffing issue": he consulted with his colleagues over and over, "weighing" and "sifting" their opinions and comparing them with his own.

One day as I was returning from the forest carrying my cross-country skis with my "tongue hanging out," I ran into Andrei Mikhailovich who was taking a walk along Mal'tzev Street. "Oh, you're just the person I need! I want to appoint R. as the head of a laboratory — what do you think; could he handle it?" We then discussed the strengths and weaknesses of my work comrade for almost a whole hour (by which time I was numb with cold). At the time I was naturally flattered that the director took me into his confidence. Later on I came to understand that, for Andrei Mikhailovich, I had been just one of his "check points," from which he would construct "the solution."

Here is another striking example of Budker's "staffing method." Andrei Mikhailovich recommended hiring Sh., a senior design engineer, into the Institute of Nuclear Physics from a neighboring institute. This caused an animated discussion to flare up around the Round Table, in which I also participated. Professor G.I. Dimov, who knew about the candidate's previous work, gave him a negative review and strongly objected to hiring him. I supported Dimov because not long before this a friend of mine, a fellow alpinist, who worked at this other institute and whose opinion I trusted, had given me the same evaluation. I must point out that the Scientific Council backed us up rather strongly — the case was clearly dubious. Andrei Mikhailovich continued to try to persuade the Council members, insisting on his recommendation. Gradually it began to seem that he might be right. The discussion ended — at any rate, the director alone would make the decision. Two days later, however, it turned out that Andrei Mikhailovich had rejected his own recommendation. Our collective opinion functioned as an indicator, and the vigorous discussion was necessary for its operation.

Andrei Mikhailovich treated his students with attention and thoughtfulness. He had a special "nose" for talented youth, and whenever he managed to discover a talented young person he always considered it to be a great success and good fortune. In the 1960's at Novosibirsk State University, I led the physics seminars for Andrei Mikhailovich's course. He was too busy to work with his assistants, and so we were absolutely independent within the limits of the syllabus, as developed by him in general outline. However, Andrei Mikhailovich thought that he should show up at the examinations. Usually he interrogated two or three students (one of them a female, as a rule) in order to "gauge" the course and then left. At one such examination, he entered the lecture hall, saw me, and asked, "Let me quiz someone who is bright." Naturally, my reaction in this situation was straightforward: I "yanked out" V. Parkhomchuk, a student in the class whose answers I always used to check my own. Andrei Mikhailovich talked with him for more than half an hour, and from the expression on their faces (already then Parkhomchik was well-known for being imperturbable) it was impossible to know how things were going. At last the examiner handed to the student his grade book, turned to me, and said in a low voice with obvious satisfaction, "This one has a real brain." About 20 years later this student made a decisive contribution to implementing one of Budker's most brilliant ideas — that of electron cooling. And all these years Andrei Mikhailovich kept an eye on him, along with many others who now determine the image of the Institute.

Andrei Mikhailovich was wonderfully gifted as a story teller. When he was at his best, one could listen to him for hours. But one day I had the opportunity to see that Budker could also listen. It happened at his apartment in Moscow. I had come on a business trip and for some reason was summoned to see the director. Almost immediately after me entered A.I. Alikhany'an, a long-time friend of Budker.[1] It was dinner time, and at the table Artem Isaakovich began to talk. The conversation began whirling: Germany in 1945 and the Nuclear Commission, the search for the German uranium project, Robert (Bob!) Wilson and the cellist M. Rostropovich — and on and on. He talked and talked, and Andrei Mikhailovich ... listened (not to mention all the others). I glanced at the director: on his face there was look of admiration for the speaker who had been a participant in these far-off events of romantic past history. Budker listened in silence for almost a whole hour! Never before had I seen such a thing, nor ever since.

Now, I will touch on Budker as a leader. His principles and his methods were original. Also, he could be resolute, he had the ability to take risks at appropriate times, and he could give up one line of work and take up another. During my first years at the Institute, I had the opportunity, under the leadership of B.V. Chirikov, to work on a device for creating the very same stabilized beam from which my "road to Siberia" had started. The work was going not so badly, although a little slowly: the specialized betatron B-3 produced an electron beam with a current of 300 amperes (which corresponds to 3×10^{13} particles, still the record for such accelerators), and further progress seemed possible. But Andrei Mikhailovich was already completely "possessed" by electron cooling. There was insufficient impetus for the work we were doing, and the future of our subject was unclear. So, in 1967, he made up his mind to terminate the experiments and to transform the group. This principle of Budker not "to become stuck on pieces of iron" (which most researchers traditionally suffer from doing) functioned more than once in the life of the Institute.

Two years later, preparations for the experiments on electron cooling of protons were in full swing. An electron beam with the required parameters had been obtained, and the design of the experimental facility was nearing completion. The drawings were brought to Andrei Mikhailovich for discussion. He looked at them for a long time and then completely routed us: a multi-layer solenoid will not produce the

---

[1] Editors' note: A.I. Alikhany'an (1908-1978) was a corresponding member of the U.S.S.R. Academy of Sciences and recipient of the Lenin and U.S.S.R. State Prizes.

required field with anything "in reserve"; the electrostatic system for beam deflection has the potential to create problems with breakdown; and on and on. A.N. Skrinsky, B.M. Smirnov, and I — the main originators of this version — at first vigorously defended ourselves. One of the main arguments in our favor was that the working drawings were almost ready for starting up construction. But Andrei Mikhailovich firmly insisted, "Don't spare the paper, or else you will have to throw away the iron!" S.G. Popov, who was then participating in this work, supported him. We yielded, beginning all over again. Thus was born the EPOCHA device, which went on to have long and productive use and whose scheme was copied by other laboratories around the world.

Next for mention is his commitment to principles. In scientific circles people talk a lot about notorious instances of coauthorship and about "bosses" who add their names to the papers of their subordinates. I received a lesson on this subject from Andrei Mikhailovich while preparing one of my first publications — a preprint about starting up a high-current betatron. The name of A.M. Budker was also included in the list of authors. When he signed the paper, he apparently noticed some doubt in my eyes. His intuition in these kinds of matters was extraordinary: "Well, do you have any doubts about my participation? After all, who invented this little magnet for you?" He showed me a picture of an entry magnet with a very original design. Thus, even though the initial idea for of the accelerator was wholly his, Andrei Mikhailovich offered a particular proof that was extremely clear and convincing for a young scientist. I remember the pleasure with which he repeated the saying that "authorship is as different from coauthorship, as singing is from sniffling."[2]

Are energetic impatience and reasonable prudence incompatible? When the experiments on electron cooling of protons in the NAP-M device began in 1973, almost every day Andrei Mikhailovich asked how the work was going. At first the work went poorly: the proton beam persisted in not wanting to travel together with the electron beam. Andrei Mikhailovich even summoned one of the plasma theorists to help, but eventually we managed it by ourselves. The effect turned out to be rather interesting. Ion pumps were installed in the region of cooling. Ions from the residual gas were accelerated by their electric field

---

[2]Trans. note: In Russian, when one constructs the artificial word "co-singing" (by analogy with "co-authorship"), one obtains the actual word that means "sniffling."

and neutralized by collisions; these entered the electron beam, which ionized them and confined them with its field. As a result, the electron beam became charged up to a positive potential, of the order of the potential at the pumps. (It was V.V. Parkhomchuk who understood this.) Afterwards, N.S. Dikansky insisted on rebuilding the vacuum chamber, which, although it necessitated halting the experiments, turned out to be the right decision: in a few months, the first effects of cooling were observed. At the time we were working around the clock, and I recall that one of the first convincing plots was obtained by the Sunday night shift on which I worked. We waited impatiently until 12 o' clock noon (the time when the Scientific Council met), and finally I took the plot to the Round Table. Yes! At last we had "caught" the effect! Andrei Mikhailovich, however, looked at the plot, smiled skeptically...and advised us not to be in too big a hurry. In fact, within a few more days we were able to obtain the next, more substantial, evidence: the "attraction" of protons by the electrons, when the protons changed their energies (and their orbits) under the influence of the electron beam.

Much has been said about the physical style of thinking. A visual lesson that I had on this subject was also connected with Andrei Mikhailovich. During one of the discussions in which the director was participating, I made a mistake in estimating the ionization losses and multiple scattering of a particle in a medium. "How can you make such a mistake?" Andrei Mikhailovich immediately responded. "One case involves the interaction with electrons and hence goes as Z, whereas the other case involves the nucleus and hence goes as the square of Z." With this example he showed me that a physicist must not "cram" formulas into his head, but should understand how to picture the effect, and then the formulas will be remembered by themselves.

For me, however, the most outstanding feature of Andrei Mikhailovich's personality was his extraordinarily nonstereotypical way of thinking, his ability to discern something new in commonplace ideas, judgments, and even events. There was a well-known anecdote about a doctor who was interested "whether the patient had perspired before dying." Andrei Mikhailovich considered this story to be an ode to a specialist: the doctor wants to be thought of as having made the proper decision, for if the patient perspired, the medicine was "right" — in spite of the fatal finale! In general it was typical for Andrei Mikhailovich to have an interest in "applied linguistics." I remember how at the Round Table, in all seriousness he discussed the difference between the Russian and Ukrainian words for "being in love"; he thought that the latter perhaps seemed more "earthy," closer to "making love."

Budker had originality of perception, the ability to separate the original from the commonplace. In the early 1960's, Andrei Mikhailovich came to Novosibirsk shortly after having visited the rocket design center of S. P. Korolyov.[3] He took delight in telling us about the "Vostok" spacecraft, how he admired the exquisite construction of the rocket and its beautiful shape, the elegance of which emphasized that the design was correct. Then, suddenly he became melancholic and said, "But the landing capsule was apparently built by other people — very rough work; it only lacks a steam radiator." Immediately he regaled us with a humorous story concerning how the design department for steam locomotives had been ordered to design the first diesel locomotive: in the overall sketch of the machine, as proposed by the designers, there flaunting itself at the front was — a steam boiler. I've never heard anything that better describes stereotypical thinking. As for the landing capsule, many years later I was fortunate enough to be able to visit Star City[4] and there see a copy of it (the original is now on display in Moscow at the U.S.S.R. Exhibition of National Economic Achievements). I immediately recalled Andrei Mikhailovich's comments, because what I saw made the same impression on me.

Budker was a wonderful lecturer. To this day his students recall with pleasure the brilliant style of exposition and the originality and freshness of ideas that distinguished Budker's lectures. This was also typical of his presentations at seminars and at conferences. For us young physicists, his assistants at Novosibirsk University in the 1960's, Budker's personality was surrounded by the halo of being a famous scientist, before whom everyone was a bit timid and concerning whose authority no one could have any doubts. One day, however, when he was already working on electron cooling, Andrei Mikhailovich made me see the Great Budker's halo in another light, when he told us the following episode. The episode occurred in Paris, in 1966, when Budker gave one of the first reports abroad about the idea of electron cooling.

---

[3]Trans. note: S.P. Korolyov (1907-1966) was the leader of the Soviet space rocket program.

[4]Trans. note: Star City is the training center for cosmonauts, near Moscow.

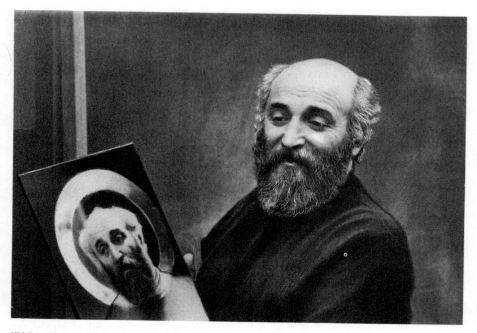

"Who thought this up?" (1975).

Questions followed his outstanding presentation, and suddenly someone in the audience asked if it might be possible for the cooled beam to be "killed" by the recombination of heavy particles with the electron beam. Nowadays, of course, we know that this is a relatively weak effect, but at that time "we had not yet thought about it." Andrei Mikhailovich said that the question stunned him — had he really missed an effect that negated the method? He had not even had time to answer when the French physicist Professor H. Bruck[5] stood up and began to ask the next question — in his typically lengthy and tedious manner. (Andrei Mikhailovich called him "boring Bruck," and from that time forth all the INP scientists used this combination to refer to this author of a well-known textbook on accelerators.) While Bruck was asking his question and Andrei Mikhailovich was answering him (fortunately, the question turned out to be an easy one), he had enough time to grasp

[5]Trans. note: Henri Bruck was a professor at the Institute National des Sciences et Techniques Nucleaires, in Saclay, France.

that, for the cooling of antiprotons, which is what his presentation was about, recombination would have absolutely no effect — since electrons, which have the same sign of charge, would be repelled from antiprotons. At that moment it was very important to come up with a completely precise answer that would leave not even a shadow of doubt concerning the effectiveness of the method.

Listening to Andrei Mikhailovich tell this episode, the thought came to me that even this absolutely self-confident man (or so he seemed outwardly) was subject to serious doubts about his own correctness. To my mind, this sort of self-criticism, or, more precisely, critical self-evaluation of one's work and ideas and suggestions, is an integral feature of a real scientist. I also thought that the situation was similar to running out of time in a chess match. (More than once in my own experience I have had opportunity to perceive how typical this is for lectures and presentations.) And, Andrei Mikhailovich could come up with a "grandmaster move."

He had a very keen sense of humor and a great appreciation for jokes and pranks. When someone played a joke on him successfully, he was perhaps even more delighted than those who had thought up and perpetrated the "funny stuff." Andrei Mikhailovich's favorite funny stories and jokes are still in "active service" nowadays.

Once, in the early 1960's, he ran into me on the stairs. "How are you?" he asked. "Well, we just launched the injector," I answered. His comeback was immediate: "How? As a satellite or as the Soviet agriculture?"[6] How could I respond?

And, of course, Budker was a dreamer. He remained a dreamer to his last days. What incredible concepts were born at the INP — social plans, as well as physics plans. There was a period when Andrei Mikhailovich was energetically developing plans to create a branch of the Institute in Sochi or in the area of Krasnaya Polyana.[7] "You see," he said at a Round Table meeting, after an appointment with the mayor of the Sochi city council, "they have a problem with employment of the youth. They have no serious industry, just fun-in-the-sun mentality and easy earnings in the summer. We could suggest to them a factory for producing commercial accelerators — "clean" industry that requires a

[6]Trans. note: The Russian verb "to launch" (a satellite) can also mean "to neglect" (a farm).

[7]Trans. note: Sochi and Krasnaya Polyana are popular resort towns in the southern part of Russia.

high level of skill." Just as in his other projects and actions, he was above all a Citizen.

The last book he read at the hospital (before he died) was a large guidebook I had brought him about the Caucasus coast.

Budker's Principles

S.G. Popov

The fact that Andrei Mikhailovich Budker was a scientist of great stature with broad interests is well-known. It suffices to list some of the physics problems to which Andrei Mikhailovich made decisive contributions: relativistic electron beams, the "magnetic bottle," colliding beams, electron cooling.... In its own area, each of these works had a cardinal influence on international scientific developments. A perusal of Budker's collected works is sufficient to rediscover how creative he was scientifically, and to perceive him as a theoretical physicist as well as an experimentalist and an engineer who found daring and unexpected solutions to problems. It was not for nothing that Andrei Mikhailovich often recalled with pride the fact that L.D. Landau called him a "relativity engineer." Hence it is unnecessary to say anything special about this aspect of Andrei Mikhailovich's activity — his works speak for themselves. To me it seems more important to fill in details about the other aspects of Andrei Mikhailovich's personality as a scientific figure — about his approach to management issues and about his interrelationships with colleagues.

These aspects of A.M. Budker's activity were exhibited during the formation of the Institute of Nuclear Physics of the Siberian Division of the U.S.S.R. Academy of Sciences, which he did at the invitation of M.A. Lavrentyev and with the recommendation of I.V. Kurchatov.

Naturally, Andrei Mikhailovich's scientific ideas were central to this new entity; however, his organizational principles and his approach to selecting people with whom to work were also very important.

It is now possible to say that his principles have passed the test of time. The Institute for Nuclear Physics (INP), which has had great scientific and technical successes, is an organization that has a distinctly unique style of work and collegial microclimate. Andrei Mikhailovich said that from the very beginning he considered the organizing of the Institute to be an extremely important and difficult scientific problem. He analyzed the experiences of other groups that had operated successfully or unsuccessfully and formulated for himself certain principles of administration, which he followed later on. It is a real pity that no complete description of them exists. However, I would like to emphasize some of these principles that I especially remember.

Andrei Mikhailovich was always wary of becoming surrounded by a narrow circle of assistants from whom he would get his information about what was going on in the Institute. He assumed that even if these assistants were the most decent of people and had the best of intentions, their information would be narrow and one-sided, and reactions to events within the Institute would become hackneyed. Therefore, Andrei Mikhailovich established the rule of very widely discussing issues in which he was interested, and he also thought it necessary to acquaint himself with different points of view, spending a lot of time, for instance, talking with certain people to whom he had no personal attraction. Budker considered scientists to be the most important people at a scientific research institute. And not only did he think so, but in every conceivable way he enhanced the authority of the scientists within the Institute. He always emphasized that the Scientific Council — the Round Table — is the highest "legislative body" of the Institute. The word "Round" carried more than a mere geometrical meaning. Among the leading scientists, the members of the Scientific Council, Andrei Mikhailovich cultivated a "director's frame of mind" approach to the solution of problems, such that the criterion for a correct decision, a criterion that unified everyone, is that such a decision is the one that is best from a global, all-institute point of view.

The decisions of the Round Table were adopted unanimously, although frequently they were arrived at after long and painful arguments. As a result, a common opinion was worked out, and even those members of the Council who were unpersuaded and who reserved their own opinions thought it necessary to vote for the common opinion. In those infrequent cases when Andrei Mikhailovich was unable to convince the Round Table, the decision was not made, even though as the director of the Institute he could have made it. "At a scientific institute, the scientists must never yield the power to administrative staff," said Andrei Mikhailovich; "however benevolent and well-intentioned the staff might be, it can bring orderliness and provide supplies and so forth, but it cannot provide a creative atmosphere." It seems to me that Andrei Mikhailovich solved the problem of giving the Round Table and scientists priority over administrative and physical plant staff, on the one hand, by distributing some of the administrative responsibilities on a voluntary basis among these same members of the Round Table and, on the other hand, by appropriately educating the administrative and physical plant staff members, in particular through the participation of staff leaders in sessions of the Scientific Council or in the Administrative Council that was formed later, in which laboratory heads were participants. As the Institute grew larger, there appeared many young talented scientists, but they were "closed off" in their everyday activities by the scientists with more experience and by the laboratory heads. In order to obtain information directly from them, so he could form his own opinion about the new generation of scientists and have an influence on their upbringing, Andrei Mikhailovich created so-called sections of the Scientific Council for each of the different scientific areas.

A.M. had extremely keen intuition: he discerned the smallest nuances in the reactions of the person with whom he was talking. With all his tactfulness, however, in one matter he was unyielding: namely, his habit of publicly promoting, in a way that was sometimes offensive to others, the reputation of someone whom he viewed as being talented. He did this in his "sermons" while discussing scientific issues, as well as in frank one-to-one conversations with potentially "jealous" persons. In the end it turned out to be beneficial not only for the "health" of the entire Institute, but also for individual scientists who began to understand more correctly their place in a scientific group and who, consequently, began to spend their energies in creative activity, rather than in unproductive emotions. Of course, this style of association cannot be prescribed in a recipe. It could happen only because of the presence of Andrei Mikhailovich's indisputable scientific authority and (I'm not afraid of using these words) wisdom about life.

In particular, I wish to point out another of Budker's unique solutions at the Institute for an important scientific-administrative problem — the problem of how to combine basic and applied work at the Institute. Since most of the scientific research at the Institute is, as a rule, done with the use of the most modern techniques and technology, sometimes with the development of novel technologies, the application of this potential to applied objectives — to investigations in adjacent fields of science, to the national economy — is often very promising. We know that Andrei Mikhailovich won permission to use money earned from introducing new technologies into industry for supporting the development of the mainline basic research, for awarding bonuses to scientists, and so on. He established a "quota" — 25% of the Institute's resources would be devoted to applied work. Thus, two problems were solved: on the one hand, researchers become interested in attracting support for their main work, and on the other hand, a highly skilled labor force is involved in ancillary applied work. In capitalistic countries, companies are known to pay considerably bigger salaries in order to attract highly qualified experts. It is now clear that the Institute's way of solving this problem has already proved its value. But not everyone knows that Andrei Mikhailovich worked on applied problems not just for mercenary reasons — mercenary from the point of view of raising support for the mainline research programs. Exaggerating slightly, he said that "doing scientific research is always satisfying to one's curiosity. Nowadays this is very expensive, and the cost is borne by society. Therefore, whenever you have an opportunity, you should repay your debt."

The first time I met Budker was on my first day of work, February 2, 1959, at that time still at the Laboratory for New Acceleration Methods, in the Institute of Atomic Energy. I must confess that my initial impressions of the Institute were disappointing: shabby premises, dirty floors whose tiles were broken here and there, sloppy experimental facilities, with wires hanging all over. Against this unattractive background, the director turned out to be bald-headed, shaven (A.M. grew a beard much later, in the early 1970's, which gave him a well-known "biblical" appearance), and not very handsome, in my opinion. But as soon as he began to speak, everything external became irrelevant. Not only was Andrei Mikhailovich a capable speaker who enjoyed talking, he also knew what, why, and how to speak. His "sermons" were never without a point — either new scientific ideas were formulated and sharpened, or issues of ethics or administration, important for the group, were discussed and promoted.

I now consider that I was very fortunate to be assigned to work at the Institute of Nuclear Physics from its very beginning; it was also very fortunate that Andrei Mikhailovich Budker, who had such a determinative influence on the personal development of many now-famous scientists, was the organizer and the director of this Institute. He was someone who could rightfully be called Teacher.

His Enthusiasm

I.B. Khriplovich

I can hardly believe that so many years have already passed since the death of Andrei Mikhailovich. Sometimes it seems to me that I saw him only yesterday, so colorful and sharp are the recollections. To the end of his life, Andrei Mikhailovich retained a wonderful wisdom as a human being and a rare gift as a physicist. But usually my memories are not about the gray-bearded man who was already marked with the stamp of a grave illness, as may be seen in his photographs during his last years. Instead I remember a thick-set robust fellow, self-confident and energetic. This is how I saw Andrei Mikhailovich for the first time, in the summer of 1959, at the Kiev Conference on High-Energy Physics. When I saw him, I immediately guessed (he was not wearing a name badge) that this was the Budker about whom I had heard so much at the time.

"The main thing is to do everything with enthusiasm. It tremendously graces life." These were not Budker's words, but those of Landau. But it is hardly possible to imagine a more enthusiastic person than Andrei Mikhailovich. Thanks to his enthusiasm, he in fact managed to live several lives in the uncompleted six decades that fate allotted to him. A keen physicist, an inventor who "bubbled over" with ideas, the creator and leader of a large institute — Budker's achievements in each of these activities would have been more than sufficient for a long, brilliant human life.

Andrei Mikhailovich was constantly attracted by the really big, difficult problems — stabilized beams, controlled thermonuclear fusion, colliding beams, electron cooling. He was far from being successful in everything, but one point, I think, is beyond doubt. If ever someone were to be able to handle such problems under such conditions, it would certainly not be the sober sceptics who often (and sometimes not without reason) criticized Andrei Mikhailovich, but rather an enthusiast like Budker.

More than once I recollected his words: "There are many wrong solutions to a problem, but only one right solution. Therefore, you can find it only if you really want to solve the problem."

Budker's cheerfulness was amazing: "The life of an optimist is preferable, for he is happy twice: when he plans his work and again when he succeeds. A pessimist can only be happy at most once: if he succeeds."

His pride was sometimes a bit naive: "I tolerate only one kind of stealing in science: when someone steals my ideas."

He was hostile to any kind of formalization: "The evaluation of scientific activity cannot be formalized. As soon as a new formal criterion for evaluation is instituted, a way to circumvent it is immediately found." Also: "The profession of theorist does not amount merely to addition and multiplication." He made a derisive reference to a certain theorist, calling him a "circus-dog mathematician."[1] One day he noticed these words of Steklov written on our blackboard: "Intuition supercedes any kind of logic." He was extremely pleased, but did not neglect to remark, "You don't even understand how correct those words are."

Occasionally Andrei Mikhailovich was very blunt, but one could answer him in the same way, without arousing any imperiousness in his nature. Sometimes he felt hurt, like a child, but one of his rules was

---

[1] Trans. note: A circus-dog mathematician is a dog trained to bark the answers to mathematical problems.

"Never take revenge." This rule did not apply only to his opponents in a verbal give-and-take.

I must acknowledge that when I became Budker's post-graduate student in 1959, our relationship could not be called unclouded at all. I occupied myself without much interest with the accelerator problems that Andrei Mikhailovich formulated, because I was attracted to another field of physics. Budker's reaction to my lack of zeal (or enthusiasm, to return to my theme) was sometimes stormy. It was only many years later, when I had my own post-graduate students, that I could appreciate the tolerance with which Andrei Mikhailovich had treated me, a stubborn youngster. I hope that I have at least partially digested this lesson of kindness.

However, the fact that Budker was able to create a large and very strong scientific institute cannot be explained merely by his outstanding personal qualities. There was one more reason for his success: he considered it absolutely necessary to foster that thing which could be conventionally called "the spirit of the institute." Andrei Mikhailovich was a real leader — resolute, firm, sometimes even harsh. However, even after he had made up his own mind, he spared neither effort nor time in order to convince the staff and the Scientific Council of the correctness of this decision, and he usually succeeded. Still, there were also cases when, heeding the opinions of others, he changed his mind. Consequently, we did not merely give some thought to the questions under discussion, but also felt ourselves to be real participants in solving the fate of the Institute.

Hence, it is not fortuitous that so many leaders of large scientific organizations of various kinds came out from the INP — a fact that half-jokingly, half-seriously, was a matter of pride to the Institute.

The older I become, the more I begin to understand how fortunate I was to have even a simple acquaintance with someone like Andrei Mikhailovich. Only now am I able to realize how correct he was in some of his judgments and evaluations. Here one cannot but recall Mark Twain's words, which I heard once from Budker: "Strangely enough, the older I become, for some reason the wiser my father becomes."

Ideas Ahead of Their Time

W. Jentschke

I remember meeting Professor Andrei Mikhailovich Budker at the CERN Symposium on high-energy accelerators and pion physics in 1956 at Geneva. There, the subject of space-charge-stabilized relativistic electron beams, presented by Budker and Naumov, aroused much interest and led to intensive critical studies at many laboratories. In the following decades I met Budker and his co-workers at many international conferences and informal meetings in the Soviet Union, Western Europe, and the United States. In all discussions with him, I was impressed by his novel and daring ideas. As the director of the Institute of Nuclear Physics in Akademgorodok, where he had moved in 1957, he established a very individual and independent style of designing and building accelerators and, particularly, electron-positron storage rings.

Budker, with his powerful personality, spoke up his mind. Talking about physicists and their accelerators, he told me of his conviction that, for research at the frontier of physics, physicists should build their own equipment. Whenever possible the components of the accelerator should be built in their own workshop. He thoroughly disliked scaling up established accelerator designs to ever larger sizes.

In the laboratory there was an air of intense creativity and there were many technologically challenging developments, but final success was not always achieved, because the performance of some of the advanced components fell short of expectations. His developments were often ahead of their time.

For all these reasons, a visit to his institute was a fascinating experience.

After the high-energy conference in Tbilisi in 1976, I visited his institute in Akademgorodok following his invitation. I spent several days there and had a very pleasant visit with Budker, his wife, and his son at their home. During a boat ride on the nearby lake we were discussing the possibilities of increased collaboration between our laboratories in the accelerator arts and particle physics.

Touring the laboratory I saw the electron-positron colliding beam storage rings VEPP-1 to VEPP-4. Important contribution to elementary particle physics were made with VEPP-2 and its high-luminosity version VEPP-2M, especially in the determination of the precise masses and decay widths of vector mesons. These experiments were done by a small group of high-energy physicists led by Professor Sidorov. In recent time, with VEPP-4, which can reach $14\,\text{GeV}\ e^- e^+$ collision energy, precise mass values of J/ψ, ψ', and Y, Y' and Y'' masses were measured, with the use of the resonance depolarization method, invented many years ago at his institute.

Continuing my visit through the laboratory, I saw Budker's factory for mass producing low-energy accelerators that could be sold to customers at home and abroad. From the gain of this sale, Budker had additional funds available for his research program.

I saw many unusual features in the laboratory: gas-jet ionization to achieve charge exchange injection into minute ironless synchrotrons; special magnets to give very short focal lengths at targets to yield intense antiparticle beams; and one of the most imaginative innovations that Budker originated, "electron cooling" for proton or antiproton beam stacking.

His idea of electron cooling, suggested in 1966, is to allow an electron beam to travel in the same direction and at almost the same velocity as the proton beam. The electrons exchange transverse momenta with the protons by Coulomb scattering. The net effect is that the volume of phase space occupied by the proton beam is reduced, making the beam denser. I was very much impressed when I could see the effect, which was demonstrated to me by Professor Skrinsky on a small storage ring called NAP-M.

The main result was that an 85 MeV proton beam could be cooled in 80 ms, reducing the beam diameter from 1 cm initially to 0.5 mm after cooling. The energy spread is reduced to less than 10^{-5} and the angular spread to less than 5×10^{-5}.

Very recent experiments done at Novosibirsk yielded a cooling time of 40 ms, equilibrium states with beam diameter of 100 micrometers, and a relative width of momenta of $10^{-6}(\Delta p/p)$. The final states are equivalent to proton beam temperatures of a few degrees Kelvin.

The new technique of beam cooling (electron cooling and, in addition, stochastic cooling invented by van der Meer of CERN) proves to be the most important advance in accelerator- and beam-technology for many years.

After visiting the Institute, we sat down around the famous Round Table for discussion. There, many excellent scientists were sitting around us, who shared with him responsibilty for major decisions. Budker said to me that there is often disagreement at this table. But his people told me that he normally could convince everybody to try brave, new ideas again.

It is very gratifying that this spirit still goes on until today.

I Appreciated Our Friendship

G.K. O'Neill

I met Andrei Budker for the first time in 1960 at the International Conference on High-Energy Physics in Rochester. Before that, in 1956 and 1959, I had presented reports on storage rings at the conferences in Geneva, and Andrei found me in Rochester in order to discuss this subject. Already then I was strongly impressed with his great energy, activity, sense of humor, and interest in life. Having learned through the interpreter that I was 33 years old, he remarked, "The age of Christ at the cross," which I thought was amusing for a Soviet citizen to say, all the more because he was Jewish.

We next met in 1963 in Dubna. During the conference Andrei suddenly had the idea to invite a small group of physicists (three people, as I recall) to Novosibirsk. With his characteristic energy and impetuousness, he went immediately from words to action. We boarded a Tu-104 airplane and in a few hours were in Siberia. The airport in Novosibirsk was small and modest, but only two years later when I again visited Novosibirsk there was a huge airport with marble floors and modern furniture in the waiting room. As it turned out, only a few weeks before us, the first foreign visitors — a group of mathematicians — had visited Akademgorodok.

Andrei was very proud to show us his Institute. An electron storage ring was in the process of being constructed here, and an electron-positron storage ring was under development. We saw the huge workshop where every detailed part of the devices created at the Institute was made.

It was in the summer, and I especially remember our walks along the broad streets of Akademgorodok and excursions on the Ob' Sea. On board a cutter I got into an animated friendly discussion with Andrei about the relative advantages of the Soviet and American political systems. The atmosphere was like that in an athletic competition. There was much laughter, and Andrei's colleagues who gathered in the narrow cabin of the ship encouraged us on.

My next visit to the Institute took place at the end of 1965, where Andrei organized a special conference on storage rings. Between sessions of the meeting, we relaxed by going cross-country skiing. Our Princeton-Stanford group in those days had run into many instabilities in our electron-electron storage ring, which were eventually overcome in theoretical and experimental studies during the 1960's. One of the best theoretical works on colliding beam instabilities was presented at this conference by A.N. Skrinsky. In the evenings, limited only by the need for translation, Andrei exchanged humorous stories and jokes with his old friend Bruno Touschek, who now has also passed away. Ludmila, Andrei's wife, belonged to a younger generation, and I remember her liveliness, sense of humor, and irony when referring to furniture in Andrei's house that was very expensive but too old-fashioned in her opinion. In 1966 Andrei and Ludmila visited the United States, and we met in Palo Alto and Princeton. During breaks in the scientific program, they both enjoyed going to shops and restaurants in Stanford, and we spent the weekend at the small seaside town of Carmel. There were lots of jokes and fun. Andrei told stories about his gymnastic feats as a youth. To show that he had not yet completely lost his former skills, he did a handstand on my lawn.

If I had to try to share my impressions of Andrei Budker, this wonderful and outstanding person, I would speak about his energy and intensity, his sense of humor and inexhaustible fountain of eloquence, and his strong feeling of independence. And, probably, it is first of all necessary to speak about his pride. He was very proud of his country and of the fact that he had created the Institute of Nuclear Physics, where thousands of people now work. He was especially proud of the talented and creative youths whom he had involved in research and who continue working at the Institute. Every moment that I spent with Andrei was wonderful, delightful, and useful. I appreciated his cordiality and friendship. He was one of the greatest people whom I have been fortunate to know.

"Cards on the Table"

N.S. Dikansky

At the end of June 1977, I met with Andrei Mikhailovich in Moscow. He did not feel very well. He was lying in bed. It was about five in the evening. The telephone was on a bedside stand, near the pillow. No one was calling at that time; in Novosibirsk they had finished the day's work, and in Moscow the workday was coming to an end. But Andrei Mikhailovich, just as he needed to breathe, needed someone to talk to.

We started talking about electron cooling. Andrei Mikhailovich began to tell about how the idea behind this method had been conceived. In the winter of 1960, Budker flew from Moscow to Novosibirsk every week to give lectures on physics at the University and to inspect the progress of the construction of the Institute. Once, he arrived in Novosibirsk with S.N. Rodionov (who now works at the Space Research Institute in Moscow). They stayed at the "Centralnaya" Hotel in Novosibirsk, since there was still no hotel in Akademgorodok. It was close to eleven in the evening, but they did not feel like going to sleep, due to the time difference. Slava had brought Agatha Christie's mystery novel *Cards on the Table*. The plot was spellbinding, and Slava began to read aloud, simultaneously translating from English into Russian. For an hour the events of the mystery unfolded, but it was impossible to understand the plot yet, and the murderer was still unknown. Slava's throat became dry, so he poured himself a glass of water. At that moment Andrei Mikhailovich's thoughts went back to working in their usual direction: the phase space volume of lightweight particles can be compressed since there is an energy loss, that is, by radiative friction. How could one overcome the constraint of Liouville's theorem if it is not electrons, but instead heavy particles — protons or nuclei circulating in a storage ring? Ionization losses occur when a particle moves through matter. By colliding with electrons in a material, protons give their momentum to the electrons, and a rather large drag force arises. Here, it seemed, is the necessary dissipation mechanism. In fact, the drag force is opposite in direction to the total velocity of a particle. With the average loss compensated, it is possible to prevent any escape from the equilibrium orbits; however, transverse momentum is not restored. In this way, the angular spread will be reduced.

Andrei Mikhailovich expounded his thoughts to Slava. Slava heard him out but immediately objected that protons would scatter on nuclei. Considerable momentum appears due to the birth of π-mesons, created during proton collisions with nuclei. Therefore the equilibrium beam size would be very large. This is a real obstacle. How can one get around it? Again, this is like a mystery, but who is the "criminal" and where is the logic behind the interrelationships of the characters in the plot?... And what if one removes nuclei and leaves only electrons, then the good features would still be the same. Take a cloud of electrons; protons move through it, and there is a drag. But is this the answer? No, for is it really possible to make the density of free electrons be the same as in solid matter? If there is no compensation for space charge, electrons without nuclei will fly off under the forces of the self-consistent fields, even though the density of electrons in electron beams is 10^8 to 10^7 particles per cubic centimeter, that is, 16 orders of magnitude less than in a solid.

No, we don't have an idea yet! Instead, Agatha Christie-like mysteries arise. Nothing is clear. His thoughts carried him away. What if one takes not an electron gas at rest, but an electron stream in a stabilized beam? It is possible to force electrons to move in external focusing fields, in which case they do not fly off. But, all the same, the density is low. How can the effect be improved? Reduce the relative velocities? The Rutherford cross section has a strongly increasing dependence for small relative velocities — the fourth power! If one makes the relative velocity of protons and electrons 10^{-3} to 10^{-4} times that of the mean velocity, it is possible to increase the cross section by 12 or even 16 orders of magnitude. There, finally, is the answer!...

The missing orders of magnitude in the density are thus compensated for. His thoughts started to click, and an overall design began to manifest itself. In a section of the ring of a proton accelerator, protons and electrons move in the same direction with equal mean velocities. The picture becomes even more familiar if we go into the coordinate system moving with the mean velocity of the particles. Then we have an ordinary two-component plasma of electrons and protons. The protons and electrons exchange energy through collisions. Temperature relaxation should occur — equilibration of the temperatures. The electrons will be colder, since equal transverse velocities implies that their temperature will be 2000 times lower than that of the protons. It was already clear to Budker who the "murderer" was. A project was being born. Morning came. It was time to go to Akademgorodok, and the car arrived. All the way there they discussed the new idea.

During his lecture Andrei Mikhailovich did not explain things clearly. He made blunders and when he finally became too muddled in his derivation of the wave equation for a bifilar line, he asked Slava, sitting next to the door, to help find his mistake. Slava turned deep red, tried to say something, and began to stammer. Then he admitted that he also did not understand. He was embarrassed when the students looked at him and saw that he was holding a book with a brightly colored cover. The lecture was obviously not a success....

At the time, listening to that lecture, naturally I myself had not suspected that on that night had been born the method for electron cooling, which was to play a decisive role not only in the development of high-energy physics, but also in my own destiny as well as that of my colleagues.

He Preached the Cult of Work

R.A. Salimov

During the first years of its formation, the Institute of Nuclear Physics had an urgent need for a fast influx of young scientists. In order to achieve this goal, a university and a physics-mathematics high school were set up in Akademgorodok, with the active participation of Andrei Mikhailovich. However, he did not wait until the first physicists graduated from the university (I think that waiting was something intolerable for Andrei Mikhailovich's energetic nature), but invented "long-term training work." The scheme for this training work was as follows: Students who had finished three years at any of the leading graduate schools in our country could compete to be selected for an appointment to begin work immediately as laboratory assistants at the Institute. During this period of work, they would continue to be enrolled as students at their respective universities. Members of the Institute would teach them specialized subjects, and all other classes would be conducted at Novosibirsk State University. In accordance with these policies, about 15 people from universities in Leningrad and Moscow and the Tomsk Polytechnic Institute were chosen in 1961. Besides Andrei Mikhailovich, this group was "nursed" by some who are now academicians — S.T. Belyaev and R.Z. Sagdeev. In those days our Institute was still small (about 300 people), and Andrei Mikhailovich knew almost everyone by sight, including us, the laboratory assistants. On the occasions when he met us, he was usually interested in our work, which had an enormous stimulating influence on us.

Andrei Mikhailovich's talks to the Institute members also had a profound educational significance. He would come out with reports on a variety of occasions: ceremonial meetings, planning meetings, and seminars. The audiences on these occasions also varied, from scientists to workers. In his reports, so it seems to me, he formulated his ideas about what is and what is not natural for a person. For example, Andrei Mikhailovich advocated the cult of work. At one of the meetings (whose subject I don't remember) in the early 1960's, he suggested to an audience of about 200 to 300 people that nothing is more important for a man than to work. Nothing — not women, nor sports, nor collecting things, nor any other hobby — could give a man such inner satisfaction and self-confidence, as could success in labor, in work.

Even when he spoke to a general audience, there was the impression that he was confiding something from deep inside himself, almost from his innermost being. When the thesis that work is foremost becomes a conviction, a person's lifestyle is shaped accordingly. I am sure that Andrei Mikhailovich was able to inculcate this thesis into many of the young members of the Institute, for it became the basis for the special spirit that is called the INP spirit. Among the leading members of the Institute, no one was involved in sports in a big way, and, for a long time, no one bothered to acquire summer cottages, gardens, or cars for themselves. Of this fact Andrei Mikhailovich was proud.

The Main Thesis of His School

E.M. Trakhtenberg

Formally I cannot consider myself to be Andrei Mikhailovich's disciple. I started to work at the Institute for Nuclear Physics in 1962, after I had already had six years of experience working in the machine-tool construction industry, with my "own" machine-tool to my name. Nevertheless, working with Andrei Mikhailovich for almost 15 years was not only a gift of fortune, but it also taught me a great many things. Not only I, but also all of the engineer-designers who worked closely with A.M. Budker would probably say the same thing.

Andrei Mikhailovich had the same enthusiasm for each new project as he had had for the very first one. In the preliminary stage, discussions continued without interruption for days. Andrei Mikhailovich involved in these discussions a very large circle of people, everyone concerned with the subject. A future device would first be "talked out" for several months — all the main problems and difficulties and the ways and specific means to overcome them would be brought out in the open. For me, these discussions were Budker's "school." It was an unusual school. Andrei Mikhailovich was such a fast thinker and reacted to the participants' remarks so quickly that it was rather difficult, especially at first, to follow him; after two or three hours of such discussions, as a rule I felt completely exhausted. These discussions were similar to a single-actor theater production: Andrei Mikhailovich surprising us, captivating us, commanding us, yet teaching us at the same time. I could formulate the main thesis of his school as something like this: "God does not exist!" We are ourselves creators; everything depends upon us; we can do anything; we just need to discover, not to dismiss; we should not try to disprove his, Budker's, unexpected ideas, even if they sometimes seemed crazy, but find ways to carry them out. Andrei Mikhailovich was the one from whom I first understood and even got a "gut feeling" for the notion that all of physics is an indivisible, interdependent, and self-consistent organism. Proceeding from the most general premises, Andrei Mikhailovich could very quickly deduce any formula, except for the constant coefficients. He was a physicist in the entire sense of the word, a physicist with tremendous breadth, which immediately became obvious, even to non-experts. Andrei Mikhailovich was proud that L.D. Landau had called him a "relativity engineer," because for Budker, the engineering to implement his physics ideas was of vital importance. One might argue about what kind of engineer Andrei Mikhailovich was. Perhaps he made too many decisions "emotionally"; perhaps his wish that each new device be technically different from preceding ones and be marked with the seal of his originality was unnecessary. Perhaps it might have been better to create something that was easier and more conservative, with large margins for error. Budker, however, in both technical matters as well as in physics, did not "ground" his imagination; instead, he "soared" and thereby attracted other scientists. "We must always be enthusiastic" was one of his mottos. "We must never forget how difficult it is for a physics institute like ours to survive in Siberia. Our steady-state equilibrium is not Moscow, nor Serpukhov, nor Dubna. It is Tomsk."[1]

Andrei Mikhailovich was very wary of what he called "small scientific fiefdoms" and therefore was constantly striving for large, difficult, and complex topics that would unify the various laboratories of the Institute. Whenever he saw the symptoms of this illness, he rapidly

---

[1] Trans. note: Tomsk is an old city and regional administrative center in Siberia. About 100 years ago, when a bridge for the Trans-Siberian railroad was built across the Ob' River at Novosibirsk, Tomsk declined in importance and was largely displaced by Novosibirsk.

made decisions for various administrative shifts, personnel reshuffling, transfers from one area to another. Any "separatistic" tendencies were also suppressed with a firm hand. Budker had high appreciation for talented and independent scientists, but if he thought that their behavior violated the "basic rules," he quickly took firm measures, including "surgery." The truth, however, is that almost every INP member who left the Institute went on to receive a considerable promotion.

Andrei Mikhailovich's personality was very vivid, in the most general sense of this word. He was interested in everybody and in everything. So many visitors came to our Institute and the Round Table! From national leaders and cabinet ministers, to Soviet biathlon and figure skating national teams — their spectrum was extraordinarily wide. Budker did not receive them merely "out of politeness"; rather, he was genuinely interested in them personally. I remember quite well how a big group of actors from the Mossovet Theater, about 70 people, came to visit the Institute of Nuclear Physics during their tour in Novosibirsk. In the conference hall of the Institute, Andrei Mikhailovich told them about our day-to-day activities, our problems, and the issues with which we were occupied. I do not know how to describe his presentation. It was not a speech or a report, but something else. He immediately won the attention of this unusual audience by being unpredictable, witty, and artistic. Later, while we were showing the Institute and our experimental devices to these actors, they kept on repeating, "Wow, what a director you have! " Although I cannot be sure, it seems that all the interesting people who visited Akademgorodok were invited to Budker's home to see him. He attracted people like a magnet, and not just because of his charm, wit, and erudition. Even during their initial meeting with him, people recognized the tremendous personality of this man, his total unorthodoxy and naturalness. By the way, about his "naturalness": Andrei Mikhailovich very much liked humorous anecdotes, even valuing them (the good ones, of course) as a kind of folklore that had survived into our day of 100% literacy. He knew a great many such stories and could tell them to perfection, with exquisite timing, at appropriate moments; very frequently, when arguable and occasionally painful questions were being discussed, one of his stories would be the final argument.

Budker's personal charisma was amazingly forceful. The most illustrative example is the story about how Alexander Abramovich

Nezhevenko joined our Institute. (Nezhevenko himself is also a legendary figure, about whom much should be written.) When he and Budker first became acquainted (in 1960), he was the director of the Novosibirsk Turbogenerator Factory — not only one of the biggest factories in the city, but in the whole country. Certain parts for the world's first electron storage ring to operate with colliding beams (the VEP-1 device) were being manufactured at this factory. It was natural that during his business trips to Novosibirsk (since the INP was still located in Moscow) Budker always met with Nezhevenko for discussions on how the joint work was going. According to Andrei Mikhailovich's own words, on one such occasion, while attending a planning meeting that was being led by Nezhevenko, he noticed the clarity of Nezhevenko's orders and his conductor-like control over the vast and complex workings of the factory, and he thought to himself, "It would be great if I had a deputy like this man! How marvelously he could organize and head up the machine-shop and all the engineering and technical services." Of course, it was easy to think such a thing. But exactly how Andrei Mikhailovich was able to make his idea a reality, how Nezhevenko, an experienced director who was far from being a youth, not only believed in Budker but also abruptly changed his career to become Andrei Mikhailovich's first and leading deputy — to be honest, this is a mystery to me even to this day. The force of Budker's "gravitational field" was, of course, irresistible for a normal person, but Nezhevenko was not a normal person. The Budker-Nezhevenko tandem turned out to be a brilliant team. Each had absolute confidence in the other, mutually trusting and esteeming and even liking each other. Nezhevenko immediately "shouldered" an enormous pile of problems related to the formation of the Institute: construction, setting up of the pilot production facilities, organization of all auxiliary services, personnel matters (except for the recruiting of scientists), housing, and so on. In establishing and training our engineers, from both a professional and personal point of view, Nezhevenko's role was almost as important as that of Andrei Mikhailovich.

One of the things to which Andrei Mikhailovich devoted a lot of attention was creating morale at the Institute, an atmosphere that would make everyone want to work. He may have spent no less time on these issues than he did on physics.

"A complicated question." On the right, Alexander A. Nezhevenko, Budker's deputy director (1974).

From the very beginning of the Institute's existence, he (more than anyone else) understood and tried to make everyone in the Institute realize that an atmosphere in which laboratories and people have good relationships would lead to a tenfold increase in achievements. If, on the other hand, the situation were to go out of control, these kinds of problems would thoroughly hinder the work and would leave the Institute far behind "in the dust." While Budker was still alive, we had opportunity more than once to understand (in a positive sense) how correct and wise he was; later, unfortunately, we also understood it, but in a negative sense.

At almost every meeting of the Scientific Council, he never tired of repeating that the success of your colleagues is your own success, because you also work at this Institute, and that "one must play for the person who spikes so that the whole volleyball team will win." Andrei Mikhailovich was always saying that it was unethical for one laboratory to reprove another — only the director could do this. He often dwelt upon the problems of the relationship between teacher and disciples, having deliberated about these issues in all of their dialectics (to put it grandly). It is a pity that no one recorded what he said at the Council — vivid and inimitable even when retold, Budker's utterances had a charm of their own. His main point was that although a clever and talented disciples may advance farther than his teacher, the disciple should never come out against his teacher! The most extreme form of protest by a disciple is to leave his teacher. Since all of us who worked with him and were taught by him were much younger, I think that for most of us (not just myself) such words did not seem to be terribly urgent, especially in our Institute where Budker's authority was both firm and high. After 10 or 15 years had passed, however, we understood that these were also our own problems, sometimes even the foremost of our problems.

Andrei Mikhailovich's reaction to something that he considered to be a violation of ethical principles was swift, sharp, unambiguous, and occasionally unpredictable. I wish to mention one specific case. At the World Fair in Montreal, one of our plasma devices — the UN-4, I think — was exhibited in the Soviet pavilion. Based on the success of this exhibition, the State Committee of Atomic Energy proposed awarding bonuses to those Institute members who had participated in making it. A certain amount of money was also allotted to the design department. However, because according to the order the award could be given only to those who had participated in this project, the administration of the design department decided to refuse the money altogether. At the time, there was intensive work on designing and constructing the VEPP-3 storage ring and other devices, and it was thought inexpedient to single out a group of persons only for the reason that their work had gotten to the Fair in Montreal. This decision was perhaps controversial. Those who had not received a bonus wrote a complaint to Budker and even collected the signatures of all the scientists and the machine-shop employees who had received a bonus for the UN-4. When Andrei Mikhailovich received this complaint with its more than 200 signatures, he was extremely indignant. No, not with the complaint itself, but with the number of signatures that had been gathered. "Do you really think that I will not respond to 20 signatures, but that I would to 200?" he asked at the departmental meeting that was held specially for this issue. "What makes you not believe in your own correctness and in me, your director?" This attempt to obtain their goal by any and all means, even "heavy artillery" (the 200 signatures), led to quite opposite results. The complaining division in the design department was disbanded, and its members were transferred to the staffs of the respective laboratories with whom they worked. I wish to emphasize once again that there were no professional reasons for this disbandment, only "moral and ethical reasons" — which, however, Budker considered to be no less important for work.

In general, Andrei Mikhailovich's attitude toward engineers was unusual and unconventional. On the one hand, having understood better than anyone else that modern equipment for physics experiments is very complex and that nobody would construct for us (at reasonable terms, that is) the unique devices that he had proposed, Budker always had a concern for the development of the engineering and industrial base of the Institute — exactly for this reason Alexander Abramovich Nezhevenko came to the INP. It is difficult to come up with an example of any other academic institution that had such large, diversified production capabilities, with great technological potential, as the INP had. Over a long period of time, he very carefully went about building up a highly skilled design department at the Institute. No one was admitted without a very detailed and relentless interview (with which I am quite familiar, having first been the subject of such an interview and later a participant). Persons who applied for the position of Senior Engineer or higher were necessarily interviewed by Andrei Mikhailovich himself or by one of his deputies, A.A. Naumov or A.A. Nezhevenko.

On the other hand, Budker constantly repeated that ours was a physics institute, created for physicists and organized so as to make physicists the main figures, with optimally convenient conditions for their work. The engineers always had to remember this. As he polemically sharpened this idea, he said one day: "To me, there's no difference between engineers and janitors — we keep them both only out of necessity." These were his literal words; I remember them because I was so shocked. Frequently he said that "engineers should not be given any freedom, for they are very business-like and energetic and will quickly 'walk all over' the physicists."

Budker always made the skilled engineering staff members clearly understand their task and the scope of their activities within the specific conditions of the Institute of Nuclear Physics.

Here, I indulge myself to recount one more reminiscence, of a very personal nature. In April of 1973, two theses were defended at the same time: M.M. Karliner, head of the electronics laboratory, defended his doctoral thesis, and I my candidate's thesis. Since both of us were engineers by training and by virtue of our work at the Institute, Budker could not pass over this fact. His concluding words concerned precisely this matter. I cannot quote his words exactly, but his main point was that engineers should not become physicists, and that good engineers would, as a rule, make bad physicists. To his knowledge, the sole exception was Christofilos, but he was only the exception that confirms the rule. However, when engineers do their job, they succeed and receive degrees. He finished his statement in typical Budker style: "If Cow is your name, an udder and milk you must claim."

A Wonderful Organizer

A.G. Aganbegyan

In 1982 there appeared on the counters of the "Akademkniga" bookstores an impressive-looking volume, a posthumous edition of the collected works of Andrei Mikhailovich Budker, academician and director of the Institute of Nuclear Physics, Siberian Division of the U.S.S.R. Academy of Sciences. It was impossible for a non-specialist to understand these papers: formulas, technical drawings, diagrams, all done most thoroughly. All of it deserved respect, combined with reverence. This is certainly the case, because it focuses on the most complicated problems of nuclear physics, the science that became a symbol of our century. But many things are absent in this book, things which should be told about its author, who was not only an outstanding scientist but also a wonderful organizer.

To Budker, winner of both the Lenin and the U.S.S.R. State Prizes, belongs credit for a number of radical physics ideas: the development of ways for confining a plasma in a magnetic trap, the idea for building colliding beam accelerators, the method of electron cooling, and many other things. It was said that he was a "divinely favored physicist."

However, this is not the point that attracts our attention now. Budker the physicist is inseparable from Budker the organizer.

His organization abilities had already been well-known at the Institute of Atomic Energy, where Budker had headed a laboratory before moving to Siberia. Even the legendary I.V. Kurchatov, whose name was given to the institute, called Andrei Mikhailovich by nothing less than the nickname "S.D." — "Sir Director." In this jestful way, he expressed a real esteem for the management skills of his subordinate.

The golden age of academician Budker's creative activities in science and organization occurred during the years when the Novosibirsk scientific center was formed. As one of the founders of Academgorodok, he had to be involved at the beginning with a wide variety of problems related to the organization of normal life in a new place: he participated at the meetings of the housing commission; he instituted clubs and set up the Novosibirsk House of Scientists, well-known throughout the country; he arranged the physics educational program at the university; and so on. Still, the favorite and the most famous of A.M. Budker's creations was, of course, the Institute of Nuclear Physics that he established in Siberia.

Not a physicist who visited from abroad failed to express his surprise at the miniature size of this institute, one of the world's leading centers for studies of the atomic nucleus. The smallness of the Institute is, of course, relative and can be understood only in terms of comparisons. It is difficult to imagine anyone who was more concerned about — who, it might be said, had more agony over — spending national funds than Andrei Mikhailovich.

Modern atomic physics is an incessant competition to develop capital investments rapidly. For the next breakthrough in experimentally checking theory, it is necessary to think of a project and quickly build the experimental device and perform the experiments. If one "dozes" even just once and falls behind, he need not be in a hurry anymore, for he is already outstripped. Moreover, in order to construct the facilities for scientific experiments in nuclear physics, the cost can amount to hundreds of millions of rubles, and even more. The mainstream way of development is not the least expensive; the size of present-day synchrophasotrons is larger than that of an average village, and their costs cannot be compared at all.

A meeting of the presidium of the Siberian Division of the USSR Academy of Sciences: from left to right, D. K. Belyaev, A. G. Aganbegyan, A. M. Budker, M. A. Lavrentyev, A. A. Trofimuk, and S. T. Belyaev (1973).

Under such conditions, the attempts of the Siberian "nuclearists" to compete scientifically with scientists from American research centers or from CERN in Europe seemed necessarily doomed to failure. Nevertheless, they won, and more than just once. The advantage was gained primarily not at the expense of resources, but due to novel ideas and to skillful, virtuoso management of the work.

At the time I am talking about, the U.S.S.R. Academy of Sciences operated as an institution on the basis of budget financing, and the portion of work that was financially self-supporting was not very large. The fact that the amount of budget support was not unlimited became obvious when it was distributed. This was especially true for the Siberian Division of the U.S.S.R. Academy of Sciences, whose financing came from the Russian Federation's Gosplan.[1] In which way should

---

[1] Trans. note: Gosplan is the State Planning Agency.

academician A.M. Budker, director of the Institute of Nuclear Physics, act? The prestige of the science that he represented was very high. There were about 50 institutes in the Siberian Division, and not all of them were equal. Therefore, perhaps the archeologists or biologists should give way and let the "nuclearists" have a larger "place under the sun." Someone who disregarded national interests and who was concerned only with his own narrow area would have acted exactly like this. Andrei Mikhailovich Budker, however, never went this way and, indeed, his way of thinking was very far from it. For him, such a "source" for obtaining increased support for scientific research did not exist.

Now, many years later, when we evaluate the way chosen by A.M. Budker, we can see that this was to become the policy in the establishment of an NPO,[2] a corporation formed to do both research and manufacturing. At that time, in the 1960's, however, the idea for an NPO, and even the term itself, did not exist.

Budker had to start from scratch, that is, from the selection of people with whom to work later. Here he put his stake in youth, which turned out to be completely justified. Nowadays, seeing academician R.Z. Sagdeev, who was director of the Institute for Space Research and is one of the leading Soviet physicists, and hearing him speak, one can scarcely imagine that only about 25 years ago this man came as a young physicist to a small group led by A.M. Budker, which was the nucleus of the future Institute of Nuclear Physics.

To find talented youth is only the beginning of the work. It is more difficult to manage talented people, because each of them has his own scientific ideas and interests. Academician Budker was able to unite them around one central subject for study. Without this unity, it would have been impossible to solve the great problems that were posed for the Institute. Lots of these problems had nothing in common with problems that are solved by a scientist standing at a blackboard with a piece of chalk, when he derives a formula or proves a theorem or convinces someone that he is right. Within the limits of being an academic scientific institute, it would go on to establish a powerful organization for the design and construction of projects, as well as a first-rate base for precision machine building. These purely managerial and budgetary problems were quite atypical for a scientist who continued to carefully

[2] Trans. note: NPO is the Russian acronym for Scientific-Industrial Corporation.

follow new theoretical ideas, to teach, to participate in seminars and conferences, etc.

People worked day and night, paying no attention to days off. The decision of the Presidium of the Siberian Division to prohibit working on Sundays dates precisely from this period. But this goal-oriented, focused enthusiasm paid off. The machine shops in the Institute built vacuum facilities that could pump down to pressures of 10^{-9} torr, at a time when the best factories in the country could deliver equipment that gave only 10^{-6}. A modern computer system was set up, in which each individual computer functions like a little screw in a complicated machine. An electron camera was designed and built that could take hundreds of millions of pictures in a second. In brief, it was the highest level, the upper limit of technological development. All these things were done for the sake of science, in order to provide greater opportunities for scientific experiments. The first of the original devices for obtaining colliding electron beams was only 6 meters in circumference. One of the last facilities (VEPP-4) was already as large as 300 meters.

The industrial and technical base, which is necessary for physics experiments, had been set up and had begun operating productively. At the same time A.M. Budker built at the Institute a factory for pilot production of the highest technical level, which, together with the Institute's design department, was able to fill various orders quickly and conscientiously.

It might seem that our story about the organizational talents of an enterprising Siberian physicist has come to its end. The factory looks for the most profitable orders; the Institute, of which this factory is a part, puts the profits into its general account and, thanks to this money, has the opportunity to conduct investigations larger than budget appropriations would allow. But let's not be in a hurry. Remember, as a scientific organizer, Andrei Mikhailovich Budker, a scientist whose name was world renowned, was absolutely outstanding.

The industrial base of the Institute of Nuclear Physics of the Siberian Division of the U.S.S.R. Academy of Sciences was firmly goal-oriented toward the problem areas with which the scientific laboratories were occupied. Yet, this academic institute began to produce elementary particle accelerators for industry. It actually manufactured them. Were there alternatives? Let's list them in order of increasing effort: consulting; formulating technical assignments for other, specialized institutes; producing technical documentation and transmitting it for subsequent elaboration or use; building experimental samples to be shown at exhibitions. It was even possible not to be entrepreneurial at all, but just to be satisfied with the support derived from the budget, following the advice to "cut out one's coat according to one's cloth."

Instead, however, machines were manufactured and delivered to industrial enterprises. These had higher requirements for safety: the people who would work with the devices were not scientists. It was necessary to train the future operators, since no one else could be responsible for adjustments and oversight while the machine's operation was being mastered. We should keep in mind that all this was going on even while the main subject of the Institute and the personal interest of its director was to understand the innermost mysteries of nature. But perhaps precisely because these were the problems before the Institute is what made its director choose to take not the easiest possible path.

At the time when the first industrial accelerators began to be mastered, prospects for the development of radiation technology were not definite. The purpose of the first accelerator was apparently the solution of an important, although isolated, technical problem: how to make cable insulation more durable and heat-resistant. Laboratory experiments showed that the plastic part of the cable casing, when irradiated, had significantly improved characteristics and was able to withstand high temperatures. However, a table-top demonstration in the laboratory is one thing; a large-scale technological process is quite another. Looking back, we can now say with certainty that the fate of radiation technology in Soviet industry would have been much less rosy without A.M. Budker. Nowadays, the strengthening of cable casing is only one of many problems that are solved with the help of radiation technology. Others include the disinsectization of food stuffs (the first such experiments on the disinsectization of grain were carried out at the personal initiative of Andrei Mikhailovich) and the problem of sewage disinfection, which was to become an active area of development after his death. There is a large number of highly varied technical problems, with scores of useful applications.

The industrial accelerators made at the INP — powerful devices that are able to function as radiation sources — are in use at many industrial enterprises in the Soviet Union and are also being sold abroad.

The income from their sales defrays more than half of the expenses for the fundamental research of the Institute.

Any economist should be envious of the organizational and financial ideas of A.M. Budker. For example, he suggested setting up at the Institute special certificates, something like internal transfer money, that would entitle the research laboratories to pay for production expenses. The concept that the Institute should be internally self-supporting gained a firm visible basis. Long before that, the idea had already been systematically inculcated in the minds of the scientists that the main pursuit of the Institute was still fundamental research.

A.M. Budker's organizational talent included, as a natural part, the ability to work collaboratively and make collective decisions. Let's imagine the situation at that time. Here was a famous scientist with ideas, who knew what to do and how to do it. Around him were young people, over whom he had great authority. These were perfect conditions for giving orders and watching to see how they are carried out.

But Andrei Mikhailovich's aspiration was to avoid this style of leadership, even in small details and in outward aspects. He put in an order for a big round table (now well-known) with a dark-colored glass top. He did not want to distinguish among participants at meetings even by who sat where, and he also did not want to distinguish himself. Whatever question was discussed, he tried to use persuasion to influence the one who would do the work; he was never concerned about spending his time doing so. If needful, he would speak about the same subject five times, as often as necessary for someone to be brought to consider his assignment as his own. The results of his style of doing work have greatly affected Soviet physics.

He died unexpectedly, not even 60 years old. In such cases, the words that are usually written in obituaries are "died suddenly." It is well-known that when the leader of a scientific group leaves unexpectedly, a new one cannot always be found immediately. But thanks to A.M. Budker's style of leadership, the administration of the Institute was a united group of confederates. The tragic event of his death could not break the continuation of mutual work.

Two Recollections

D.D. Ryutov

1. Andrei Mikhailovich and Fusion

Andrei Mikhailovich actively participated in the initial stage of the work on controlled thermonuclear fusion. The most famous and impressive result of his activity during this period was the proposal of a new scheme for confining plasma, based on the use of so-called "magnetic stoppers" ("mirrors"). The idea of the "Stoppertron" is the basis nowadays for practically every kind of open trap, which is one of the most promising ways to achieve thermonuclear fusion.

According to Andrei Mikhailovich's recollections, during the first years when thermonuclear research was begun, many of the participants had the illusion that quick success would be possible, and this determined the specifically technological direction of the work. He stated that he had also paid tribute to this tendency, by doing calculations on an inductive method for extracting the energy that comes out of a thermonuclear reactor based on the Stoppertron.

Experience, however, showed that this initial optimism was unfounded. Plasmas began to present one surprise after another to the physicists, and it turned out that the behavior of a plasma could not be understood with the style of thinking based on the description of single particle motion in an electromagnetic field.

Later, Andrei Mikhailovich liked to remark that at the time, he was one of the first persons who clearly understood and then actively began to promote the idea that further progress would be impossible without investigations into the physics of plasmas, investigations that were not directly related to studies of the workings of specific thermonuclear fusion facilities. Limiting oneself to mere general appeals was not Andrei Mikhailovich's style. He began to act in the appropriate direction. In particular, while thinking about how to be able to investigate oscillations in a weakly nonequilibrium plasma conveniently, Andrei Mikhailovich proposed using thermal ionization of the vapor of alkaline metals. The experiments on alkaline plasmas with temperatures of only 2000 or 3000 degrees (whereas in a thermonuclear reactor, the temperature must be 50,000 times higher!) turned out to be really quite productive in terms of "manufacturing" new physics information and making progress in the solution of some key questions of plasma physics. At the end of the 1950's, during the establishment of the Institute of Nuclear Physics, Andrei Mikhailovich took measures to provide a basic physics style for the plasma work at the Institute. In my opinion, however, Andrei Mikhailovich himself was not greatly attracted to the detailed "rummaging in the entrails" of plasmas. For the next ten years, his obvious preference was for problems in high-energy physics and he moved away somewhat from the plasma physics community.

Young people (including myself) who entered plasma physics at the end of the 1950's and the beginning of the 1960's grew up in an environment where priorities had already been firmly assigned to pure physics investigations. Every day, interesting problems arose in connection with the "collective" nature of plasmas; new physics ideas were conceived (which later turned out to be very useful, not only in plasma physics); and arguments about plasma turbulence were in full swing. Our enthusiasm was supported by the active involvement of famous senior physicists such as L.A. Artsimovich, M.A. Leontovich, and E.K. Zavoysky in this work. My contemporaries always remember this time as a very happy time.

In the midst of all these discussions and the sparkle of beautiful theories and interpretation of nice experiments, the final goal of our work was nearly forgotten, and it was almost impolite to talk about it. I remember how one young theorist proudly declared that he did not even know the order of magnitude of the cross-section for the deuteron-triton fusion reaction, because he never needed it (and, at that time, it was the truth!).

As mentioned earlier, at that time Andrei Mikhailovich had moved away somewhat from plasma physics, although he continued to monitor the situation, being kept informed by collaborators and colleagues with whom he had previously worked. Keeping an eye on the course of events as if from the sidelines, he noticed, at the end of the 1960's, something that had apparently eluded those who were directly involved in the work, namely, that they had become far too carried away with "pure" plasma physics and that the knowledge that had been acquired was, in fact, quite enough for plasma physicists to turn to their original mission. The appropriate occasion for Andrei Mikhailovich to point this out occurred in connection with hosting the conference of the International Atomic Energy Agency in Novosibirsk in the summer of 1968. In particular, he made this statement in his report at the close of the conference:

"It seems to me that the successes that have been achieved by physicists in this field during the past years make us return to the idea of creating a thermonuclear reactor. It is not necessary for a physicist to know everything before getting down to business. It is not necessary to wait for the last button to be sewn on the greatcoat of the last soldier before starting battle."

Furthermore:

"The question is very often posed concerning how long it will take to build a thermonuclear reactor. The answer to this question is analogous to a famous story about a traveler and a wise man. Once upon a time, a traveler came to a wise man and asked how long it would take to reach the nearest town. The wise man answered, 'Go on, get going!' The traveler shrugged his shoulders in surprise, but the wise man repeated, 'Get going, and then I'll tell you.' He started off and then glanced back. The wise man thereupon said, 'Keep going and don't look back.' So the traveler went straight off. 'Now I can say how long it will take you to go,' uttered the wise man, 'because now I know how fast you go.' Only after we have considered the problem in its entirety and when we know the means and the people that are available for its solution and the attention that is paid to it, then we will be able to answer the question of how long it will take to build a thermonuclear reactor."

The published text of his report is much shorter than what Andrei Mikhailovich actually said. The point is that he spoke without a "crib-sheet," improvising, returning over and over to one or another idea, as if he were considering it from different angles. This was the first time that I heard Andrei Mikhailovich give a speech about "general matters" (not a specific scientific report), and initially I felt something like disappointment: I was accustomed to hearing much more compact and stylistically sharpened speeches, of which L.A. Artsimovich was a master, and it even seemed to me that Andrei Mikhailovich was not prepared for the report. Gradually, however, he captivated me (as he did most of the audience). Later I came to understand that Andrei Mikhailovich always behaved in this way, making no exception even for the most crucial speeches. It was quite necessary for him to reflect during his speech, verifying his thoughts and shaping them to absolute clarity. He really "thought his idea through to the end" during a speech, involving his listeners in this process. (The same thing was typical of his speeches at the Scientific Council, to which I will come back later.)

Nowadays, when we know how subsequent events transpired (indeed, how fusion began to make real progress in going from the 1960's to the 1970's), his speech seems quite natural and logical. At the time, however, it did not seem so at all, and it caused a strong internal reaction on the part of many people. It required courage to speak as Andrei Mikhailovich did.

During the same report, Budker touched on one more aspect of the problem. He stated: "The problem of thermonuclear reactions is not a usual physics problem. This is a problem that must transform society and the world. Our generation, which gave to mankind atomic energy and thermonuclear energy in the form of explosions, is responsible to humanity for solving the main problem of energy — obtaining energy from water. People are waiting for the solution of this problem. Our duty is to solve it within the lifetime of our generation, and therefore we must set out on this path."

The idea that the physicists of his generation — the generation that had created nuclear weapons — were indebted to humanity and really must solve the problem of controlled thermonuclear fusion was apparently important to Andrei Mikhailovich; he often returned to it.

At the beginning of the 1970's, in connection with the events described earlier, Andrei Mikhailovich's interest in the problem of controlled thermonuclear fusion intensified again. The question arose about selecting a specific direction of work on this problem for our Institute. Already at that time it was clear that the tokamak outstripped all other types of thermonuclear devices, and we were tempted to join this most popular trend. Such a decision would not have been unnatural for Andrei Mikhailovich, because he had formerly made an important contribution to the physics of tokamaks, having drawn attention to some of the features of plasma diffusion in such devices (from his comments later grew up the so-called "neoclassical" theory of transport processes). After some reflection, however, he came to the conclusion that it would be better for an academic institute to work on problems in a less developed area of fusion, where the emphasis would be placed on the investigatory nature of research. Consequently, it was decided to study open traps. In some of their features, these devices have significant (potential) advantages over tokamaks. However, their physics is less developed. Also, at the beginning of the 1970's, it was known that plasma escaped from open traps too rapidly through the stoppers, which could make it quite impossible to make a thermonuclear reactor based on them. The main difficulty in this case was associated with the highly "nonequilibrium" state of the plasma in a Stoppertron, which leads to plasma instability.

Andrei Mikhailovich suggested looking for a solution to this problem by greatly increasing the plasma density — to values for which the mean free path of the plasma electrons and ions would be much shorter than the length of the device. In this case, the plasma would be much closer to thermal equilibrium. However, for values of the density permitted by the mechanical strength of the vessel, the device turned out to be too long. There were some more discussions — and the outline of a solution of the problem began to take shape: it would be necessary to construct the device as a sequence of individual Stoppertrons, each with an appropriately chosen length. Thus arose the scheme for the multiple mirror trap.

A fundamentally new scheme for plasma confinement that was suggested in 1971 by Budker and his disciples. The theoretical predictions were entirely confirmed with this small device.

In the first months of 1971, when these discussions were going on, Andrei Mikhailovich did not fully recover from a heart attack and was unable to come to the Institute sometimes for a whole week. However, his desire to solve the problem was so strong that he invited us — young physicists — to his home and forgot about the time and his illness in exciting discussions with us.

The authority of Andrei Mikhailovich and the charisma of his personality were so great that the fact of his sincere interest in the problem was sufficient to distinguish it clearly in our eyes from all other problems. As a result, many scientists at the Institute became simultaneously interested in the problem of improving open traps. Therefore, it is not surprising that in the middle of the 1970's, already without the direct participation of Andrei Mikhailovich, more new schemes were suggested at the Institute for open traps, which, together with the multiple mirror system, today determine the role of the INP in fusion.

The final results of this work are not yet perfectly clear, but the existence of several lines of investigation that back up each other provides a good chance for success.

2. Andrei Mikhailovich at Meetings of the Scientific Council

The meetings of the INP Scientific Council that took place every Wednesday did not merely reduce to discussions of current issues, such as plans for work, competitive elections to open positions, and so on. Very often, when Andrei Mikhailovich was "in good spirits," the conversation would imperceptibly turn into what many people now call "sermons." They began spontaneously, often triggered by some remark of a member of the Council. Andrei Mikhailovich's favorite subject was that of relationships within a scientific group. As everyone knows, they do not always develop smoothly. Clashes between personalities and ambitions, struggles for leadership and acknowledgment of scientific priority, aspirations to self-affirmation — all these, apparently, are the unavoidable concomitants of the scientific process. If these things sprout up, work becomes difficult, especially when its goal is the solution of big problems that require the coordinated efforts of a large number of participants, which are precisely the kind of problems that comprise the base of the scientific program at the INP.

At the beginning of a conversation, Andrei Mikhailovich would seem to be thinking out loud, not having earlier prepared his words (so it appeared to me). In this case he needed the active and well-disposed participation of his listeners, their interested attention, and their objections. When too many people raised objections and the discussion became too noisy, Andrei Mikhailovich would interrupt opponents firmly. This would not cause offense, because Andrei Mikhailovich was much older than the other members of the Council (at least at the time when I was a participant) and was its undisputed leader. Besides, his goodwill never failed. Also, his very appearance, which invited comparison with that of a sage Biblical prophet, corresponded to his style of talking.

Andrei Mikhailovich liked to draw the analogy between a well-coordinated scientific group and a good family, in which each person — consciously or not — has a feeling for how much of the common resources he can claim, and in which these resources are divided up without long debates, as if occurring automatically. Andrei Mikhailovich emphasized that this sort of model for life is realistic, if the members of the group are ready to yield to each other in small things.

Frequently Andrei Mikhailovich touched upon the subject of personal responsibility. He said something to this effect: "All of us like to 'blame the authorities,' and everyone considers the 'authorities' to begin with his own direct boss, while forgetting that for many others, he himself is an 'authority.' Therefore, before one starts to cast blame, he must think about his own actions."

Another problem about which he was deeply concerned was that of the relationship between teacher and disciple. Here he held the view — which could possibly be disputed — that the disciple has no right, under any conditions, to come out against the teacher. The most extreme degree of protest permitted to a disciple is to leave the teacher.

For me, the enjoyment of my many contacts with Andrei Mikhailovich was connected with the opportunity to be touched by his life's wisdom and to be an eyewitness of how he prepared himself for making critical decisions, considering an issue from various aspects and discussing the possible and often quite unexpected consequences. It seems to me that in the solution of the problems of life he found creative satisfaction no less than he did in physics.

The Invention of the Gyrocon

V.S. Panasyuk

The situation was normal: Andrei Mikhailovich was looking for someone to listen to him, for without such a person (it did not matter who it was) the creative process gave him no pleasure.

And so we went for a walk in the evening, along the paths of the Golden Valley.[1] First he had to tell a few jokes, then an interesting story involving a woman. I was beginning to think that the entire conversation would be in the same vein, and that the walk would be casual and perhaps even relaxing. But I am seldom so fortunate, especially in such situations. Abruptly there appeared on Andrei Mikhailovich's face an expression of anxiety: the question of a powerful high-frequency power supply for linear colliding-beam accelerators was not yet solved. He immediately launched into a creative "musical improvisation," pausing briefly now and then to ask, "Do you understand?" It was both useless and unprofitable to say "No." First of all, this irritated him. Secondly,

[1] Trans. note: The "Golden Valley" is a pleasant area on the outskirts of Akademgorodok with many birch and aspen trees whose leaves turn golden in the autumn.

having to explain things that he thought were obvious stopped the creative process. Besides, it was not only unnecessary, but even detrimental, to understand his "intermediate" models, for their obviously fantasized nature generated annoyance on the part of his listeners. "I understand," I repeated almost automatically, looking forward to the end of a hard walk. Suddenly, there was an apprehensive look on Andrei Mikhailovich's face. He described a very simple idea, which we both understood. It was the idea for the gyrocon. To tell the truth, later it turned out that this type of device was already known. At the time, however, we thought otherwise.

The following day he called in Oleg Alexandrovich Nezhevenko and asked him to draft the text for a patent application. I think that he did not dare rely on me.

Exception to the Rule

H.P. Furth

Experiments in plasma physics and high-energy physics have much in common: in both cases, the central endeavor is the electromagnetic confinement and energization of charged particles; in both cases, particle behavior is dictated by a combination of classical scattering and collective phenomena. At the outset of large-scale modern plasma physics research, in the 1950's, it would have been natural to expect the emergence of a common scientific leadership — giving shape to the new fusion program on the basis of experience in high-energy physics, and at the same time applying the insights of plasma physics to the design of more advanced accelerator experiments.

Generally speaking, this constructive interaction of the two disciplines did not materialize; major figures in experimental high-energy physics have not made important contributions to experimental fusion research, and conversely. There are, however, notable exceptions to this rule — and the most notable is clearly Andrei Mikhailovich Budker. His electron-positron colliding-beam experiments alone, or his

mirror-confinement ideas alone, would have made him a leading figure in modern physics research. His integrated grasp of high-energy
physics and plasma physics concepts, combined with an extraordinarily creative mind, gave rise to a sustained output of new ideas that
have proved seminal in both fields.

The opportunity to know Gersh Budker personally did not present
itself to me until the summer of 1967, at the so-called "Shock Conference" in Novosibirsk. On this occasion, Budker and his colleagues had
gathered together a diverse group of international scientists — some
working on plasma shock waves, some on electron beams, some on
stellarators, and others on a variety of additional topics matching the
far-flung interests of the hosts. Interspersed with discussions of positron
and antiproton beam problems, astrophysical conjectures, and new insights into plasma transport in tokamaks, there were outings on the
Ob' Sea and evenings of memorable conviviality. The Shock Conference in Novosibirsk is recalled by its U.S. participants as an altogether
unique intellectual and cooperative phenomenon.

During subsequent years, I had a number of other opportunities to
meet with Budker, and was always struck anew by the breadth of his
interests, the fertility and boldness of his imagination, and the warmth
of his humanity. The news of his death saddened me deeply.

Fascination That Infected Everyone

V.V. Parkhomchuk

My first encounter with Andrei Mikhailovich occurred in 1962. He came to lecture about high-energy physics and plasma physics at the physics-mathematics summer school held in Akademgorodok, in the Novosibirsk suburbs. This summer school was attended by secondary-school children who had been selected on the basis of the results from the two test rounds of the First All-Siberian Olympiad. His lecture opened up what were for me new, completely unexplored vistas of science and made a deep impression on me. His presentation was clear and easy to understand. For instance, the advantages of the method of colliding beams were explained with the help of the illustration of two locomotives traveling in opposite directions that collide. I cannot claim to have understood all these ideas, because the first time I had ever seen a locomotive was when I came to this school, but what I did remember was his explanation of physics as cognition, an ongoing process of deepened understanding. It was clear that after finishing school and graduating from the university, one could become a direct participant in this exciting hunt for new knowledge.

My next, more regular, encounters with Andrei Mikhailovich occurred at Novosibirsk State University, where he gave lectures on mechanics. The explanations of this "relativity engineer" (he told us, not without pleasure, that he had been called this) not only clarified the most complicated problems of relativity theory, but also accustomed us students to the idea of using it for practical purposes. While attending the Institute of Nuclear Physics, I was able to witness how these ideas, which had come up at the juncture of relativity theory and plasma physics, were being implemented — for example, the work on creating a relativistic stabilized electron ring. During my examinations, I had to meet personally with A.M. Budker. The questions from the examination sheet only served as a starting point for him to vigorously probe my knowledge on a broad area of problems. Our conversation ended by his asking me where I was from and who my parents were.

In 1972, work began on the development of one of A.M. Budker's ideas, that of electron cooling, and I was fortunate to participate in it as Andrei Mikhailovich's post-graduate student. The problems and the experimental results were actively discussed not only at the Institute, but also at Andrei Mikhailovich's home. If, while he was at home, a new idea suddenly came to him or he learned about a new result, he would immediately summon all the participants in the work to his home for an enthusiastic discussion of the new ideas. He infected everyone who worked with him with his enthusiasm. When I recall these years, I can only be amazed at his gift of foreseeing something even when there seemed to be not even slight indications. As an example, I wish to cite his attitude toward the effect of the small longitudinal temperature of an electron beam, which I discovered. After I had noticed this effect, I was able to explain certain phenomena observed in the experiments, which, of course, was pleasing. Even I, however, was surprised at Andrei Mikhailovich's enthusiastic response to this notion. Later on, this effect that had not seemed so important to me gave birth to an entirely new direction in the physics of electron cooling.

In the fall of 1976, Andrei Mikhailovich was resting in the Crimea, apparently in compliance with the orders of his physicians. By that time, electron cooling had been tested experimentally; in addition, it turned out that implementing colliding proton-antiproton beams at the Institute of Nuclear Physics would interfere with the necessary development of colliding electron-positron beams. Clearly, joint work with

other institutes within the scope of the UNK program[1] in Serpukhov
was necessary. Andrei Mikhailovich was so impatient to get on with
this idea that he requested that someone be sent to the Crimea in order to develop and specifically define his recommendations. I was the
one who was sent on this strange business trip, from the Institute of
Nuclear Physics to the House of Writers in Coctebel, where Andrei
Mikhailovich was staying. I rented a small room where "wild vacation-
ers" in the Crimea usually stayed and went to Andrei Mikhailovich
every day. He, instead of being on holiday, worked on the rationale for
transforming the Serpukhov accelerator to a storage ring with colliding
proton-antiproton beams. I basically had to play the joint role of being a
devil's advocate against his fantastic ideas and serving as his secretary.
Neither megagauss magnetic fields, nor super-short bunches of particles, nor kiloampere electron currents bothered Andrei Mikhailovich.
My timid doubts only served to spur on the search to substantiate these
ideas. The result of this business trip was a 15- or 20-page manuscript
that was the basis for the proposal that the Institute of Nuclear Physics
presented at the All-Union Conference on Accelerators in 1976. I must
point out that the proposed ideas have not yet been implemented and
that their time has only now come. Andrei Mikhailovich, when he asked
for an immediate development of this proposal, realized of course that
it would not be carried out by the very next day — but it was not his
habit, if he had an idea, to delay getting on with the work.

---

[1] Editors' note: UNK is the acronym for Accelerator and Storage-Ring Complex.

He Enriched My Life

E.L. Ginzton

I am pleased to have been asked to contribute my remembrances to the book celebrating the seventieth anniversary of Dr. Budker's birthday.[1]

It is hardly necessary — perhaps impossible — to properly recognize Professor Budker's contribution to technology. Others, in addition to myself, will have written about the leadership role he played in the development of the Nuclear Physics Institute of the Siberian Division of the U.S.S.R. Academy of Sciences. His contributions to the development of colliding beam storage rings parallel in many ways similar developments at SLAC — and were thereby responsible for the many visits made to SLAC by Professor Budker and his colleagues, as well as visits to Professor Budker's laboratory by American co-workers.

It would have been easy for Professor Budker to follow the development of high-energy physics in the United States and in Europe; but that simply was not his style. He wanted to enhance the importance

---

[1] Trans. note: This statement pertains to the Russian edition.

of his work by finding novel ways of attaining the desired result. It is not surprising, then, that some of the ideas he carried out resulted in new inventions, for which he applied and received a number of U.S. patents.[2]

Professor Budker was not satisfied with his important contributions to the world of high-energy physics. He also wanted to make use of his own and his laboratory's unusual competence by applying radiation technology to commercial and industrial practice for irradiation of food — such as wheat and other grains.

He also felt that the ideas for practical application of radiation need not be limited to the U.S.S.R., but could be extended to foreign countries. By making his ideas and machines available outside of the U.S.S.R., he was able to obtain foreign currency which would enable him to equip his laboratories with foreign-made devices in a way which would not have been feasible otherwise.

The ingenious approach which Professor Budker always sought in trying to reach his goals made his contacts with foreign scientists particularly productive. Many visitors to the U.S.S.R. chose to visit him in his laboratory as a matter of prime interest. Many of us will not forget the eagerness and joyful expression he had in describing his ideas and accomplishments. I remember how the visit by the United States Academy of Sciences to Novosibirsk in 1973 brought to our group his boundless enthusiasm for his work.

I also remember in 1975 an exchange of letters with him about the practicality of my bringing my family for a visit to Novosibirsk. Not only did Professor Budker encourage us to make this trip but offered to meet me at Khabarovsk for a trip to Novosibirsk. As it happened, Professor Budker could not get away from his post but the arrangements he made for my family and me to visit Novosibirsk and his variety of activities were most productive.

One of my colleagues at Varian, Craig Nunan, participated in the U.S.S.R. National Synchrotron Radiation Conference in July of 1982. At that time, Mr. Nunan offered a toast at the banquet which I would like to repeat:

---

[2]Editors' note: G.I. Budker held 13 Certificates of Authorship in the Soviet Union and 35 foreign patents.

I wish to propose a toast to a man who is not here — yet his presence is everywhere. Shall we raise our glasses to the memory of Academician Budker?

I think this toast was fully appropriate for a gathering of accelerator physicists and engineers because Professor Budker's pervasive contribution to the field is immeasurable.

I am glad to have known Professor Budker as he enriched my life and the lives of several of my associates. I am certain that many others around the world share this feeling.

Cool Before Drinking

F.E. Mills

I first met Andrei Budker at the International Conference on Accelerators at Dubna in the summer of 1963. I had heard of his work on storage rings and wanted to compare his ideas with mine. I had, in 1961-1962, spent a year on leave from MURA at Saclay near Paris. In that year I had worked with the Orsay group on plans for a new storage ring (they were bringing ADA from Frascati to Orsay to see how high a current they could achieve with the powerful Orsay linac) to collide positrons and electrons. In the course of that work I (and at the same time Claudio Pellegrini at Frascati independently) solved the problem of how to make strong focusing electron-positron storage rings where all degrees of freedom were stably damped. This led to the e^+e^- storage ring "ACO," whose construction began when I returned to Madison, Wisconsin. Andrei invited many of us at the Dubna Conference to visit the Academic Town, and I believe perhaps 15 of us accepted. I chose to go with the early group, together with Arnold Schach of CERN and Jerry O'Neill of Princeton. We had the honor of being the first Western physicists to visit the Institute of Nuclear Physics (a group of mathematicians had preceded us as the first visiting scientists). I recall asking to have the electron-electron storage ring energy turned down to 45 MeV so that I could see the same orange-color synchrotron radiation we had in the FFAG at MURA.

Subsequently I visited INP in 1965 at the Colliding Beams Section of the International Conference on Multiparticle Physics together with Burt Richter of SLAC, Bernie Gittelman of Cornell, Gerry and Vera Fischer of CEAL, Bruno Touschek of Frascati, Yuri Orlov of Yerevan, and others. Many plans were hatched at that meeting, probably leading to SPEAR, CESR, VEPP 3 and 4, and others. My own plans for a storage ring for synchrotron radiation were already under way. I recall that the issue of one bunch versus many in a ring was debated hotly, as were all of the currently important issues, including beam-beam interaction. At that time, we learned of Andrei's interest in proton-antiproton colliders, but it seemed to be awfully distant.

In 1966 Andrei and Ludmila Budker with Sasha Skrinsky visited the MURA laboratory near Madison. While Andrei and Sasha were at MURA, my wife Joyanne took Ludmila to visit the students in the junior high school where she taught. My understanding was that it was an enjoyable experience for all, because Ludmila was so fluent in English and was willing to answer the students' questions. Andrei addressed the staff of MURA, with translation given by one of our engineers (Igor Sviatoslavsky).

Joyanne and I attended the All-Union Accelerator Conference in Moscow in 1968. We were invited to spend a week at INP, which we gladly accepted. I recall that it was the first time that I really had the time to visit all activities in the laboratory. I had the opportunity to visit the plasma laboratories of Dimov and Nesterikhin, and to have some discussions again with Chirikov and Sagdeev. At that time I was beginning to take part in the plasma confinement experiments of Kerst *et al.* at the University of Wisconsin. I was quizzed very seriously about our plans at the old MURA Laboratory (soon to be the Physical Sciences Laboratory of the University of Wisconsin) to operate the new storage ring for synchrotron radiation for solid state, atomic, and surface physics, and for chemistry experiments. I enjoyed telling the staff about how Ed Rowe had solved the longitudinal (Robinson) instability with a rudimentary feedback system. My fondest personal memory of this visit was a ride in academician Kurchatov's automobile, which

Andrei had retrieved from Moscow. I will simply say that this was very early in Andrei's career as a driver, and as he demolished a few bushes and trees, the normal driver was beside himself with concern for what was happening. Well, we all went through that at some point in our lives.

The next time I went to INP was late in 1976. We had decided to pursue proton-antiproton collisions at Fermilab, and Bob Wilson asked me to find out about the recent electron cooling results at INP, to see if we could employ it at Fermilab. I had the second task to establish a collaboration between Fermilab and INP. I spent several long, fairly formal sessions in discussion with Andrei and others. It was not completely satisfactory from his point of view, because he wanted to take part in the construction of the superconducting magnets for the Fermilab Doubler. As I suspected, Bob was not interested. We did, however, establish a collaboration in targetry, lithium lenses, cooling, and all the stuff of antiproton production and collisions, as well as other useful things in the high-energy physics program at Fermilab. Under the terms of the agreement, which I actually authored, there were many extensive visits by scientists from both laboratories. From my perspective, there was a real benefit to Fermilab. I hope that INP feels that it benefited also.

One of our regular visitors at Fermilab was Kolya Dikansky, who had so brilliantly constructed NAP-M and carried out the INP cooling experiments. When we were close to cooling at Fermilab, Kolya brought us a bottle of vodka. He had written on it simply, "Don't open without cooling."

The first cooling took place at 4:40 AM one morning. Don Young and Peter McIntyre had tuned the ring and electron system all night and I came in at 4:00 AM for the next 12 hour shift. I noticed that although the conditions favored resonance crossing, the beam loss pattern was not usual, so I asked for a few minutes to check the RF. I simply turned down the RF from several kV to 10-15 V, and after a slight frequency adjustment, observed the beam (all 100,000 protons!) cool into a tiny bucket. We immediately called Russ Huson. When Russ arrived, we and others who happened along, opened and drank the vodka according to Kolya's instructions.

It was a great pleasure for me to know and meet with Andrei Budker. I believe that he was a great creative man, who did much to introduce new ideas into our field. We all were better for knowing him.

One-Third Jewish?

N. Rostoker

Gersh Itskovich Budker (1918-1977) acquired the names Andrei Mikhailovich during the heroic era (1946-1951) of the Soviet "Manhattan Project." I.V. Kurchatov liked to rename "special" people who had difficult names. He told Budker that his names were too "Jewish" and invented better names for him.

Budker liked "Andrei Mikhailovich" and insisted that all of his wives (five in all) use this name. He was not cynical about family life. Rather, he was a romantic. His romances almost always ended in marriage. I once remarked in a toast that he had kept trying it until he had finally gotten it right. Neither he nor his last wife, Alla, were displeased by my remark.

The Institute of Nuclear Physics was founded by Budker in 1958. It now has grown to about 3000 employees. In late 1991, 14 years after his death, it became officially named the Budker Institute.

Roald Sagdeev came to the Institute in 1960. He and Budker were on very good terms and had similar tastes in humor. A typical conversation at the time might be as follows. Budker to Sagdeev: "Why did you Tatars oppress Mother Russia for 300 years?" Sagdeev to Budker (answering a question with a question, like a Jewish mother): "Why did you crucify our God?"

Budker had a very high regard for Sagdeev and helped him become an academician. By 1970 their relationship had deteriorated (I don't know exactly why), and Sagdeev returned to Moscow. In 1975 they were on good terms again, and during the period 1975-1977 Sagdeev helped Budker maintain the independence of the Institute of Nuclear Physics against the opposition of Lavrentyev and Marchuk.

Budker procured a very large round table for his conference room. Its purpose, in his mind, was to symbolize equality, with no special place for the Chairman. About 1965 he sounded like an orthodox communist, although he never became a party member; later in life he became quite skeptical in private conversations.

Budker liked to make deals. A typical case was Lia Melnikova, who came from Askhabad in the Turkmenian Republic in 1974 to buy an accelerator for her studies of radiation damage of semiconductors. Budker introduced her to me as this "Indian woman." She was Jewish, spoke some English, and looked Mideastern. She was accompanied by her daughter and a great many giant melons from home. She had cash for an accelerator, was treated like a queen, and conferred a large melon on her favorite subjects. I believe that she bought an accelerator. Budker talked like an orthodox communist, but had the instincts of a capitalist and developed a significant part of the support of the Institute from deals within the Soviet Union. He also was interested in deals with the United States. I remember that about this time Alan Kolb was at a conference in Novosibirsk. At the traditional conference picnic, a tent was erected, and Budker and Kolb sat inside, talking deals all afternoon. They both emerged at the end of the day full of vodka. I do not think that any of the deals materialized. Possibly they forgot what they were.

Budker's Jewish origins were well-known and were expressed by him on numerous occasions. His favorite question when interviewing prospective scientists in the early years of the Institute was: Is it possible to be one-third Jewish? He had no apparent religious convictions. He liked stories related to religion. About 1968 he came back from a trip to India. He had acquired a consuming interest in yoga.

Andrei Budker with Alan Kolb and others at a picnic by the Ob' River during the 1967 Shock Wave Conference in Novosibirsk.

With great animation he told the remarkable story about a Christian sect in India that believed that Jesus Christ had not died on the cross, but had used yoga to fool the Romans into thinking he was dead. As a result, he was removed from the cross while still alive, recovered, and then traveled to India where he lived another 20 years. There is even a grave marker in Kashmir that is claimed to be his last resting place.

I myself met Budker for the first time at the Many Body Conference in Novosibirsk in 1964. Between 1965 and 1977 I visited Novosibirsk five times and each time met him either in his office or socially at some dinner party or picnic.

The last thing I learned about Budker's remarkable sense of humor was the following. After his heart attacks, his heart bothered him when making love. He consulted his doctor, who said — try to do it without being so emotional. Thereafter he frequently used this phrase when there were arguments around his Round Table. I don't know if he was aware of it, but many people in the Institute were cognizant of the origin of this remark, and it usually lightened up the arguments far more than Budker expected.

Author's acknowledgment: These reminiscences are partially based on conversations with Dmitri Ryutov and Andrei Prokopenko.

As I Remember Him

R.Z. Sagdeev

I was a graduating senior in theoretical physics at Moscow University when I met Budker for the first time. He was looking for fresh recruits to join him. But in 1955 his name was known only within the narrow circle inside the highly classified nuclear establishment and did not impress us students. On my first encounter with Andrei Budker, I hadn't paid much attention to his invitation. I was much too preoccupied with my plan to join Landau's group. The glory of Landau had so overshadowed everyone that I had no intention of learning more about Budker and what he was doing in physics. Later, when I became a member of the Kurchatov Institute, I discovered what an unusual man and scientist Budker was. During my time at Kurchatov, he led a rather modest-sized team that developed new ideas for particle accelerators. My friends in fusion would tell me that his innovative ideas in that field were bordering on science fiction. Soon, I discovered his impact on the controlled fusion research I was doing.

A couple of years after the pioneering work of Sakharov and Tamm (on the tokamak), Andrei Budker suggested an alternative approach under the code name "corkatron." A few years later we learned that the Americans had a similar design named the mirror machine, suggested by Richard Post. Kurchatov obviously liked Andrei Budker and his ideas. The corkatron project was given a budget and eventually a large group of experimental physicists and engineers, led by Igor Golovin, one of Kurchatov's close assistants. In one of the laboratories in the Institute they tried to build an actual plasma machine to implement Budker's idea.

It happened that the very first plasma instability I discovered (in a joint paper with Leonid Rudakov) was precisely for this device, Budker's corkatron. After identifying this particular phenomenon, I was asked to deliver a talk at the Artsimovich seminar. It was understood that the presence of plasma instability in Budker's invention threatened the very viability of the project. The huge room was overcrowded. Such an attendance, as I discovered somewhat later, could be explained by the psychological drive so characteristic of the scientific community. They were expecting a clash, and there was something bullish in the overall desire to witness it. Fortunately, in science, usually it is a fight between different ideas and interpretations. Nevertheless, as Landau would say, people "sensed the smell of frying."

Andrei Budker, of course, was invited. Everyone expected an exciting duel between the two of us. I delivered the speech and my arguments were rather straightforward.

When I had finished my talk and answered a number of questions, mostly of a technical character, everyone expected a violent reaction from Andrei Budker. Instead, he stood up and generously congratulated me for "a very interesting contribution." Obviously, many of the attendees were disappointed. However, Budker and I, from that time on, became good friends.

In the coming months and years I had a chance to learn much more about him. Budker had gone directly from a student's desk to the front line of the Great Patriotic War. After demobilization at the war's end, he came back to Moscow and asked for an appointment with Lev Landau. The great man liked the young soldier, but was concerned whether a few years in the trenches had not suppressed his skills. In an interview he found that Budker still possessed a physicist's thinking and intuition. However, quite expectedly, he had lost his mathematical skills for dealing with formulas. Lev Landau gave him good advice:

"Young man, take your time. After recovering your math technique, come again."

Alas, Andrei Budker couldn't wait. He had come to Moscow with his young wife and a baby son. He desperately needed a job, and at least a room in a communal apartment. At one of the meetings with demobilized comrades-in-arms, some of the physicists there told him about a newly opened "mailbox" on the outskirts of Moscow that was currently recruiting scientists. That was how Budker approached Igor Kurchatov and was accepted.

Within a few years, Budker had developed into both an excellent theorist and an engineering and design genius. "The relativistic engineer," a characterization suggested by someone, fit him perfectly. At the same time, he didn't like to spend hours and hours in cumbersome mathematical exercises doing routine calculations. He had a great flair for making quick estimates — as physicists would describe it, on the backs of envelopes or paper napkins. He was always "ever-ready" for the eternal intellectual adventure.

Budker never returned to Landau. Their styles and interests by this time were drastically different. However, he always had respect and admiration for "Dau," often commenting, "Landau does not think with the regular brain of a common man — he uses his spinal cord." It was the highest compliment.

My next and rather dramatic encounter with Budker took place in the spring of 1961, when I myself was building a small team of younger disciples. While doing some part-time teaching at the Moscow Energy Institute, I discovered a few promising students and invited them to join a seminar. Among the first attendees were Alec Galeev and Volodya Zakharov. From time to time, I would give them different problems to solve. Volodya was especially bright in mathematical problems. Projecting into the future, I could see Volodya as a member of my team.

Suddenly, one evening, he came rushing to see me in a state of complete desperation.

"Roald, I am in big trouble and don't know what to do. There is a very high risk that I might be drafted by the army."

"But that is impossible!" I exclaimed. "All students have a special deferment."

"Not anymore," he said. "I was kicked out of the Institute."

I couldn't believe it. He was considered one of the brightest seniors. His expulsion, however, was not related to his academic performance. Apparently he had become a hero and, at the same time, the victim of a rather bizarre story. One night he had been awakened in his room, in the students' dormitory, by his girlfriend. She was in a state of great panic. She told him that another student, after a party, had tried to seduce her. In a state of extreme rage and agitation, Volodya immediately jumped up, found the guy, and beat him severely.

The good news was that apparently theoretical physicists are not free from such earthly feelings as passion, love, and extreme jealousy. However, the bad news came after an investigation had been carried out by the authorities, with the participation of the Komsomol. Volodya Zakharov was expelled from the Institute for "exceeding reasonable limits" in the use of brute force. Losing his place in the Institute would mean that he would automatically lose his temporary student's permit to live in Moscow.

In a fraction of a second, I figured out that the only person who could save him was Andrei Budker. He was still in Moscow, but his Institute of Nuclear Physics in Akademgorodok near Novosibirsk had already been decided upon. That gave a him a legal chance to recruit this unlucky young man. The next morning, I rushed to see Budker.

"Dear Andrei Mikhailovich, I have to tell you that one of my brightest young students has gotten into big trouble, and only you can help him."

Budker invited Volodya to tell him the story. After hearing the sordid tale, Budker gave his own assessment: "Volodya, I believe that you behaved like a real man. For a man who was, and maybe still is, in love, what you did was quite understandable. But if you are also an intelligent man, you should have found a different solution. An intelligent person should never resolve a dispute with brute force. Now let's talk about your future."

That is how Volodya Zakharov was sent as a "junior technician" to Akademgorodok in Siberia. While I was still in Moscow, I had to accept the responsibility of providing assignments to keep this "technician" busy performing all the tasks that would be typical of a theoretical physicist. That's how I was finally — simply by force of circumstances — linked to that distant campus in Siberia. Soon I too became a member of Budker's Institute, as the head of the new plasma theory group. Volodya was automatically included in it, along with a few other young fellows. One of them, Alec Galeev, had been Zakharov's classmate at the Moscow Energy Institute. He voluntarily abandoned Moscow, terminating his senior year, in order to join my theory group in Novosibirsk.

Andrei Budker, from the very beginning, knew the secret of the right balance between science and administration. He combined the abilities not only of a brilliant scientist and an energetic administrator, but also of a fine psychologist. At critical moments, he knew which buttons to push to obtain better support from the top. He also knew who at the top would have a genuine fascination for science and be ready to serve as an audience for a passionate discourse on his recent ideas. This is how Budker achieved a special place with Igor Kurchatov. Budker knew, on the other hand, that he would have less success with sarcastic Lev Artsimovich, who would express doubts about his science fiction-type suggestions, requesting more details, proofs, and calculations.

In his frequent deliberations on science, people, and society, Budker himself characterized Artsimovich as a brilliant critical thinker: "Artsimovich is so bright he can immediately find a flaw in the arguments and the theories of anyone. Probably even Einstein would have been unable to come up with his discoveries if he had been permanently exposed to Artsimovich's criticisms."

Once, in the infancy of Akademgorodok, a group of my contemporaries created an organization called the Council of Young Scientists. I was not present and thus had no idea until later that it was I, *in absentia*, who had been elected as the president of the council. My new responsibilities carried with them a rather vague and undefined role. I went to see Andrei Budker to ask for his advice. It came immediately. "Roald, if you need the advice of old men" — at that time, he probably was no more than 44 years old, but according to the charter of the council, the age of Jesus Christ was the limit — "then listen to me. Scientists are divided not according to age — the young versus the old —, but according to their actual value — the smart versus the mediocre. This is a law of value that works perfectly on the intellectual market."

Within the Institute, Budker's style was to nourish a team of like-minded scientists, and for a rather long time I was one of them. He shared with us all of his ideas, his hopes, and his doubts. In return, he requested complete dedication, even self-sacrifice.

Our alma mater, the Institute of Nuclear Physics, in Akademgorodok near Novosibirsk, was built on the principle of collective brainstorming. Budker had ordered the workshop of the institute to design and produce a rather fancy round table which could comfortably accommodate a couple dozen young scientists and their director. The very shape of the table symbolized the spirit of democracy in scientific discussion. Budker himself liked to compare it with the round table at which King Arthur conferred with his brave knights. Andrei Budker was the true personification of an extrovert — he liked to think aloud. For that reason, we were lucky to be not only witnesses, but also a part of his inner "kitchen" while he was cooking a new and quite often bizarre feast.

Budker needed criticism and suggestions in order to elaborate new ideas jointly. But sometimes he behaved like a grown-up kid, being offended by the remarks of skeptics. "It is so difficult to find something interesting and new, but instead of helping me to develop fledgling ideas in their infancy, you guys immediately look for a way to attack them."

Sometimes one of us would use a different argument: for example, "Andrei Mikhailovich, I would be ready to agree with your idea, but it seems to me to be so natural, almost simple. I cannot imagine why someone else did not conceive of such an idea long ago. It must be known already."

Budker would reply, "Your reaction reminds me of a chat between Europeans at a nineteenth century continental party: 'England is an island? Impossible. If it were so, it would have been known since long ago!'"

At the same time, Budker was very proud of his boys. The first years of the institute were tremendously successful. By 1963, we had already built our first particle collider, an early prototype of the future great machines of high-energy physics.

The creative atmosphere in the institute was a direct result of the continuous effort and attention Budker paid to the issue of hiring people and working with them every day. I do not remember a single instance when somebody was appointed to one of the scientific teams without first being interviewed by Budker himself. That was important for the stability of the institute. Those who were finally selected had the huge luck not only to work in a most exciting place, but also to be taken care of by the director, Budker himself. He helped in every respect, from the allocation of apartments where to live, to finding a job for someone's wife, to getting little kids into kindergarten.

Quite often Budker would take the role of confidant in family matters, a kind of self-assigned family counselor. He knew that the breakup of a family in a very small town could be quite detrimental, and he might lose the best of his team. In return, the boys had full respect and confidence in him. They would come looking for important advice, even on quite intimate matters.

— "Andrei Mikhailovich, I need your help. I am on the verge of an important decision in my life — to marry or not to marry."

— "If you are asking such a question, you are obviously not ready to marry. Do it only when you would feel you couldn't not marry."

The net result of his policy and his tireless efforts to communicate daily with team-members prepared us well. We were ready for major undertakings and anxious to compete with the best scientific teams in our country and abroad. Budker liked to say with pride, "We have a superiority of bright heads." No one could counter his statement. All of us accepted the rules of the game in which Budker was the enlightened tsar. The rest of us, in moments of greatest dissent, were "loyal opposition to the king."

Those who eventually crossed this line or who brought ideas or proposals that were not part of the riverbed of Budker's interests had better leave, go away, build their own ecological niches or their own scientific empires "elsewhere."

That is eventually what happened to me some years later. Although it was a painful change at the time, now I know that there is nothing wrong with such an approach. Everyone who worked with Andrei Budker knew him as a great scientist and someone who was generous and caring. This is how I will always remember him. I will never forget one of our long debates at the round table when he was sketching out the rational scenario for the further development of Akademgorodok. To do so would require the building of many more stores, boutiques, and kindergartens. Suddenly Budker added, "But we shouldn't forget about having a cemetery." The company of young men immediately burst into laughter.

"Don't laugh," he said. "But be prepared not only to live here, but to die here, too."

In 1977, I went back to Akademgorodok from Moscow as an official envoy of the U.S.S.R. Academy of Sciences and also as a friend of Budker's to deliver the official eulogy at his funeral. I bid him a last farewell at that very cemetery.

Excerpted Recollections

B. Richter, R.F. Schwitters, L.C. Teng, K. Strauch, and C.M. Braams

From "The Next Generation of Electron-Positron Colliding-Beam Machines" by B. Richter

The development of the colliding-beam technique for high-energy physics began slightly more than 20 years ago with the work of three groups of physicists — one in the United States, one in the U.S.S.R., and one in Italy. These pioneering groups were the Princeton/Stanford collaboration in the U.S. (W.C. Barber, B. Gittelman, G. K. O'Neill, and B. Richter); the Novosibirsk group in the U.S.S.R. (G.I. Budker and his colleagues); and the Frascati group in Italy (F. Amman, B. Touschek, and colleagues). From these first small machines, as our understanding of the behavior of the intense stored beams evolved, have come even larger colliding-beam facilities.

The recollections in this chapter are excerpts from articles originally published in *Problemy Fiziki Vysokikh Energii i Upravlyaemogo Termoyadernogo Sinteza* (Topics in High Energy Physics and Controlled Thermonuclear Fusion), edited by S.T. Belyaev (Nauka, Moscow, 1981).

311

The success of the technique is indicated by the fact that all new machines now under construction are colliding-beam machines.

Professor Budker played an important role in these developments. I first met him in 1965 and continued to see him every few years thereafter. My first impression remained my impression throughout — he was a man with enormous vitality, great charm, and originality. His death in 1977 was a loss to the accelerator physics community and to high-energy physics. Were he still alive, he too would be thinking about working on the next generation of machines....

From "Beam Polarization in High-Energy e^+e^- Storage Rings" by R.F. Schwitters

Academician G.I. Budker, whose life and work we are honoring here, had interests in and made significant contributions to an extraordinarily wide range of areas in physics. One theme that runs throughout his productive career is the application of classical and "well-known" concepts in physics to entirely new problems and phenomena. One example of this is beam polarization in high-energy e^+e^- storage rings.

In 1964, Sokolov and Ternov showed that under certain conditions, synchrotron radiation with spin-flip should lead to a gradual build-up of spin polarization of electrons and positrons circulating in a storage ring. Since then, physicists at the Institute of Nuclear Physics in Novosibirsk, under Budker's leadership have played major roles in both the theoretical and experimental understanding of this phenomenon and in exploiting it as a tool for studying high-energy physics and quantum electrodynamics.

From "Electron Cooling and its Application to Antiproton-Proton Colliding Beams at Fermilab" by L.C. Teng

I will begin by reading a letter from Prof. Robert Wilson, director of the Fermi Laboratory, to Professor Skrinsky and his colleagues:

Dear Friends,

I deeply regret not being able to participate in the memorial for our former colleague and dear friend, Gersh Budker. The memory of Budker is kept alive every day at Fermilab because our cooling project is building on Budker's ideas and on his and his colleagues' great achievements at Novosibirsk. Budker dreamed of collaboration. A proper memorial to Budker, in my mind, is to realize a vigorous collaboration between you at Novosibirsk and us at Fermilab. I pledge my best effort to that end.

Warm regards, R.R. Wilson

...Actually all over the United States his memories are being kept alive by various things he has invented and conceived. The electron cooling at Fermilab, the Gyrocon project which is being pursued now at Los Alamos, and the H^- charge-exchange injection which is being used at Argonne and at Fermilab, are all examples of the achievements of Professor Budker and his colleagues at Novosibirsk. In addition, of course, we have the many colliding-beam projects in various laboratories in the United States.

From "The Crystal Ball Detector" by K. Strauch

I am greatly honored to have been invited to join you, my colleagues and friends from the Novosibirsk Institute of Nuclear Physics, at this memorial seminar for Gersh Itskovich, Andrei Budker, at the commemoration of his sixtieth birthday, and at the celebration of the twentieth anniversary of the Institute which he founded.

If there is one characteristic of Andrei Budker the man, which stood out above all others, it was his vitality. And so he would have liked his memorial seminar to be combined with a celebration of his Institute, a look at both its past and its future.

Much was said last night about Andrei Budker's scientific and technology achievements. I will not attempt to repeat this list. Instead I want to read to you the program of the Memorial session for Andrei Budker being held this very day in Washington, D.C., as part of the spring meeting of the American Physical Society. The program covers several areas of physics to which Budker had made very significant contributions. Because of this memorial session in Washington, Professors Panofsky, the organizer, and Richter, one of the speakers, are unable to be here today. They have asked me to express their deep regret.

A similar conflict has prevented Professor Weisskopf from joining us today. He has given me the following to read to you:

I am sorry that I cannot be present at the memorial session for Andrei Budker. The passing of Andrei Budker is a tremendous loss for physics. The loss is much greater than the loss of a productive physicist. The reasons are obvious. There are so few people in the world community who can replace him. He had a very special character that is found rarely among our colleagues. It is the combination of his deep interest in physics combined with an unusual engineering ability. What distinguished him more than everything else was his tremendous ability to get things done under difficult circumstances. We all know the hard conditions under which every high-energy physicist has to work. Nobody was so ingenious as he in his ability to overcome any kind of difficulties. The word "impossible" didn't exist for him. On the contrary, the more difficult the task, the more he was attracted to it. But there was more to him than that. The solutions that he found to solve his problems were always original, unexpected, simple, and effective. Original in a sense that only he could have thought of a solution, unexpected because everybody else was astonished that one could do it that way, simple because it was always the most direct way that looked obvious after he had achieved it, efficient because he always succeeded in transforming his ideas into realities. I am not only speaking of physics or engineering ideas, I am also thinking of problems of human collaboration, of organization of work, and of laboratory management. He did all these things different from every one else, better, cheaper, faster, and more elegantly.

He will be irreplaceable and his loss will be felt for a long, long time. There will be fewer ideas and fewer instruments around without Budker, but his spirit lives on in his students, collaborators, and friends. We will do what we can to realize some of his ideas that have not yet been made to work, and we will think of him when new physics will be found in the future, with the help of his ideas.

Victor F. Weisskopf

Andrei Budker was deeply interested in furthering scientific collaboration among all countries. He was most anxious to make the great round table international. Many visitors have come to this Institute at his invitation and they always found the experience most memorable because of the large number of new ideas being visibly transformed into hardware. The time scales sometimes had to have Lorentz factors applied to them, but the ideas were always original and fascinating, and the results most interesting and useful.

Perhaps I can give you an idea of some of the impressions of this visitor by telling you the following memories from among many others.

On my very first visit, about ten years ago, a younger colleague collected me at the Hotel and brought me up to the third floor of the Institute. Budker was waiting at the great round table, and after introductions told me that he wanted to start by giving me a personal review of weak, electromagnetic, and strong interactions, and of the place of colliding beams in the unraveling of their relationships. The younger man turned to Budker, said, "I have heard all of this many times before; I'll be back in an hour," and left with the full understanding of his boss. In how many laboratories in the U.S. would this have happened? In how many in the Soviet Union? This certainly was an excellent demonstration of the fantastic spirit that Budker installed in his Institute and which is responsible for our great confidence in its future.

The second observation concerns Budker's great influence within your country due to his achievements and personality. Unlike other Intourist bureaus that I have visited, the one at Novosibirsk always includes at least one very beautiful lady who receives the visitor with a charming smile whatever hour of the day or night he arrives. I always suspected that Budker arranged that, too, in order to foster international scientific exchanges with the Novosibirsk Institute. And from what I understand, if there is one person who can influence Intourist in a positive direction, it was Andrei Budker.

My final recollection concerns my last meeting with Budker on July 1, 1977, three days before his death. He was resting in the hospital but came to his house to take care of some business and to see me. We had a long talk in the garden about the latest discoveries in e^+e^- physics and also about some administrative problems. As so many times before, Budker's son Dima translated his father's words. As always, Budker listened carefully to his son. When Dima made the slightest mistake, his father would correct him. Both Dima and Andrei clearly know English well; I never found out who knew it better. It was wonderful to observe the great love between father and son. Budker looked and talked like a prophet from the Old Testament: he discussed the administrative problems in terms of a rather involved story full of biblical wisdom. I was so fascinated whether and how he was going to return to the point of our discussion (which he did), that I concentrated all of my attention to this and thus have very little recollection of the story itself. I shall always cherish the memory of this last talk in the beautiful Siberian birch woods with a great and wonderful man.

From "Studies of Plasma Physics in Toroidal Systems" by C.M. Braams

I remember Professor Budker as a man with a clear scientific mind and with the attitude and inspiration of an inventor. For him, thinking about purely scientific problems, and considering their applications in the sphere of production, were two aspects of the intellectual process. I remember, for example, how in his invited lecture during the Conference on Plasma Physics in Moscow, he appealed to us to concentrate our efforts on the solution of the nuclear fusion problem. In his talk at this Conference which remains stamped on our memory he demonstrated how fundamental research on plasma physics should be combined with the search for its nuclear fusion applications.

The so-called problem of interrelation between science and society must have been no problem to Budker. Contributing to the advancement of pure science and to the development of society were like the two sides of one medal to him.... I consider it a great honor to commemorate the great scientist Budker.

Information on the Authors

Abel Gezevich Aganbegyan, academician, is the rector of the Academy of National Economy. From 1967 to 1985 he was the director of the Institute of Economics and Organization for Industrial Production, of the Siberian Division of the U.S.S.R. Academy of Sciences. Thereafter he was the academician-secretary of the Economics Department of the Presidium of the U.S.S.R. Academy of Sciences until 1991.

Anatoly Petrovich Alexandrov, academician, was president of the U.S.S.R. Academy of Sciences from 1975 to 1986. He is a three-time Hero of Socialist Labor, a Lenin Prize laureate, and a four-time recipient of the U.S.S.R. State Prize. He graduated from Kiev University in 1930. He was the director of the Kurchatov Institute of Atomic Energy from 1960 until 1987. His main work has been in the areas of nuclear physics, solid state physics, and nuclear reactor design.

Lev Mitrofanovich Barkov, academician, has been a laboratory head at the Budker Institute of Nuclear Physics of the Siberian Division of the Russian Academy of Sciences since 1967. He graduated from Moscow State University in 1952. His fields of expertise are neutron physics, nuclear physics, and elementary particle physics.

Vladimir Nikolaevich Bayer, doctor of physics and mathematical sciences, is a professor and laboratory head at the Budker Institute of Nuclear Physics of the Siberian Division of the Russian Academy of Sciences. He graduated from Kiev State University in 1954. His work is in the area of elementary particle theory and high-energy physics. He has worked at the Institute of Nuclear Physics since 1959.

Spartak Timofeevich Belyaev is an academician. He graduated from the department of physics and technology of Moscow State University in 1952. From 1962 until 1978 he was the head of the theory department of the Institute of Nuclear Physics of the Siberian Division of the U.S.S.R. Academy of Sciences and, at the same time, the rector of Novosibirsk State University. At present he is the director of the Nuclear Physics Division at the Kurchatov Institute of Atomic Energy. His main work is on the theory of relativistic plasmas, many-body quantum theory, and nuclear theory.

Cornelis M. Braams was the director of the FOM Institute for Plasma Physics "Rijnhuizen" at Jutphaas (later called Nieuwegein), the Netherlands, from its founding in 1958 until his retirement in 1987. He graduated from the University of Utrecht in 1956 after a two-year stay in the nuclear physics department of the Massachusetts Institute of Technology. He became a part-time professor in plasma physics at Utrecht in 1963. He has served on a variety of international committees and boards in thermonuclear fusion research. He is currently engaged in studying the history of fusion research.

Boris Valerianovich Chirikov, academician, is head of the theoretical department of the Budker Institute of Nuclear Physics of the Siberian Division of the Russian Academy of Sciences and professor at Novosibirsk State University. He graduated from the department of physics and technology of Moscow State University in 1952. His main scientific work is in the areas of classical and quantum dynamics and statistical physics. He is one of the originators of the physics theory of chaos. He has worked at the Institute of Nuclear Physics since 1958.

Nikolai Sergeyevich Dikansky, corresponding member of the Russian Academy of Sciences, is a laboratory head at the Budker Institute of Nuclear Physics of the Siberian Division of the Russian Academy of Sciences. He was also the dean of the Physics Department of Novosibirsk State University until 1992. He graduated from Novosibirsk State University in 1964. A specialist in the physics of charged particle beams and accelerator technology and one of the originators of the electron cooling method, he has worked at the Institute of Nuclear Physics since 1964.

Gennadi Ivanovich Dimov is a corresponding member of the Russian Academy of Sciences and a laboratory head at the Budker Institute of Nuclear Physics of the Siberian Division of the Russian Academy of Sciences. He graduated from the Physico-Technical Department of Tomsk Polytechnical Institute in 1951. His main work is on charged particle accelerators and controlled thermonuclear fusion. He has worked at the Institute of Nuclear Physics since 1960.

Yakov Borisovich Fainberg is an academician of the Ukrainian Academy of Sciences and departmental head at the Kharkov Institute of Physics and Technology where he has worked since 1946. He graduated from Kharkov State University in 1940. He works in the area of plasma physics and plasma electronics and is one of the originators of collective methods of particle acceleration.

Harold P. Furth is the former director of the Plasma Physics Laboratory at Princeton University. He is a member of the U.S. National Academy of Science. His main work is in the area of plasma physics and controlled thermonuclear fusion.

Edward L. Ginzton is chairman of the scientific advisory board of Varian Associates, Inc., in the U.S.A. He was a professor of electrical engineering at Stanford University from 1947 to 1968. He is a member of the U.S. National Academy of Science and the U.S. National Academy of Engineering. He is a prominent expert in the area of powerful ultrahigh-frequency electronics and measurement technology in the UHF band.

Igor Nikolaevich Golovin is a doctor of physics and mathematical sciences and a recipient of the Lenin Prize and the U.S.S.R. State

Prize. He graduated from Moscow State University in 1936 and, from 1944, worked at Laboratory No. Two of the U.S.S.R. Academy of Sciences. From 1950 until 1958 he was first deputy to I.V. Kurchatov at Laboratory No. Two (The Institute of Atomic Energy). At present he is a section head at the Kurchatov Institute of Atomic Energy.

Willibald K. Jentschke has been director of the DESY physics center in West Germany (1959-1970), director-general of CERN (1971-1975), and a professor at Hamburg University (1956-1970 and 1976-1980). He is presently a consultant and an honorary member of the DESY directorate. His main work is in high-energy physics and accelerator technology.

Boris Borisovich Kadomtsev is an academician, recipient of the Lenin Prize and the U.S.S.R. State Prize, and an honorary member of the Swedish Academy of Science. Since 1973 he has been the director of the plasma physics division at the Kurchatov Institute of Atomic Energy. He graduated from Moscow State University in 1951. His main work is in plasma physics, controlled thermonuclear fusion, and magnetohydrodynamics.

Alexei Georgievich Khabakhpashev is a doctor of physics and mathematical sciences, a professor, and a laboratory head at the Budker Institute of Nuclear Physics of the Siberian Division of the Russian Academy of Sciences. He graduated from Moscow Institute of Engineering and Physics in 1952. His main work is in the area of elementary particle physics and in the application of nuclear physics methods in adjacent areas of science and technology. He has worked at the Institute of Nuclear Physics since 1960.

Iosif Bentzionovich Khriplovich is a doctor of physics and mathematical sciences, a professor, and chief research fellow at the Budker Institute of Nuclear Physics of the Siberian Division of the Russian Academy of Sciences. He graduated from Kiev State University in 1959. His main work is in the area of elementary particle theory. He was one of the authors of the work on parity violation in atomic transitions. He has worked at the Institute of Nuclear Physics since 1959.

Vladimir Il'ich Kogan is a theoretical physicist, candidate of physics and mathematical sciences, and senior scientist. He graduated from

Moscow Institute of Engineering and Physics in 1947. His main work is in the area of plasma physics and controlled thermonuclear fusion. At present he works at the Kurchatov Institute of Atomic Energy.

Eduard Pavlovich Kruglyakov is a corresponding member of the Russian Academy of Sciences, recipient of the U.S.S.R. State Prize, and deputy director of the Budker Institute of Nuclear Physics of the Siberian Division of the Russian Academy of Sciences. He graduated from Moscow Institute of Physics and Technology in 1958. He specializes in the area of plasma physics and controlled thermonuclear fusion. He has worked at the Institute of Nuclear Physics since 1958.

Anatoly Alexeevich Logunov is an academician, Hero of Socialist Labor, and recipient of the Lenin Prize and the U.S.S.R. State Prize. He was vice-president of the U.S.S.R. Academy of Sciences until 1991. He graduated from Moscow State University in 1951. From 1963 to 1974 he was the director of the Institute of High-Energy Physics at Serpukhov. From 1977 to 1992 he was the rector of Moscow State University. His main work is in the areas of quantum field theory and elementary particle physics.

Moisei Alexandrovich Markov is an academician, a Hero of Socialist Labor, and formerly the academician-secretary of the Division of Nuclear Physics of the U.S.S.R. Academy of Sciences. Since 1934 he has worked at the Lebedev Physics Institute of the Russian Academy of Sciences and, since 1951, also at the Joint Institute of Nuclear Research in Dubna. He graduated from Moscow State University in 1930. His main work is in the areas of quantum mechanics, classical electrodynamics, quantum field theory, gravitation theory, and neutrino physics.

Igor Nikolayevich Meshkov is a corresponding member of the Russian Academy of Sciences and a laboratory head at the Budker Institute of Nuclear Physics of the Siberian Division of the Russian Academy of Sciences. He graduated from Moscow State University in 1959. The main areas of his scientific activity are the physics of charged particle beams and accelerator technology. He was one of the originators of the electron cooling method. He has worked at the Institute of Nuclear Physics since 1959.

Arkady Beinusovich Migdal (1911-1991) was an academician and a theoretical physicist. He graduated from Leningrad State University in 1936. His main work was in the areas of atomic and nuclear physics, quantum field theory, theory of metals, and other problems of modern theoretical physics. Together with L.D. Landau, he founded a new direction in theoretical physics: the application of the methods of quantum field theory to many-body problems. He established a school of physics. He last worked at the Landau Institute of Theoretical Physics of the U.S.S.R. Academy of Sciences.

Frederick E. Mills has been director of the Midwestern Universities Research Association, professor of physics and director of the University of Wisconsin Physical Sciences Laboratory, chairman of the accelerator department at Brookhaven National Laboratory, scientist at Fermi National Accelerator Laboratory. Currently he is a senior scientist at Argonne National Laboratory. His research interests are accelerators, high-energy and plasma physics, energy loss of fast particles in motion, photoproduction of pi mesons, and advanced accelerators.

Lev Borisovich Okun' is an academician of the Russian Academy of Sciences and a laboratory head at the Institute of Theoretical and Experimental Physics in Moscow. He graduated from Moscow Institute of Engineering and Physics in 1953. His main work is in the area of theoretical elementary particle physics.

Gerard K. O'Neill (1927-1992) was a professor of physics at Princeton University from 1965 to 1985. He invented and developed the technology of storage rings for the first colliding-beam experiment at the Stanford Linear Accelerator Center. He also founded the Space Studies Institute. The main areas of his scientific work were experimental elementary particle and nuclear physics, accelerators, and space research.

Alexei Pavlovich Onuchin, doctor of physics and mathematical sciences, is a laboratory head at the Budker Institute of Nuclear Physics of the Siberian Division of the Russian Academy of Sciences. He graduated from Moscow State University in 1959. The main areas of his scientific activity are elementary particle physics and experiments on electron-positron colliding beams. He has worked at the Institute of Nuclear Physics since 1959.

Vadim Semionovich Panasyuk, doctor of technical sciences, is a recipient of the Lenin Prize and the U.S.S.R. State Prize. In 1944 he graduated from the Radio Department of the Moscow Institute of Communication Engineering. He worked at the Institute of Nuclear Physics of the Siberian Division of the U.S.S.R. Academy of Sciences from 1958 to 1968. At present he works at the Scientific Research Institute of Optical and Physics Measurements. His main work is in the area of charged particle accelerators.

Wolfgang K.G. Panofsky is an emeritus professor of physics at Stanford University. From 1961-1984 he was the director of the Stanford Linear Accelerator Center. He is a member of the U.S. National Academy of Science. The main areas of his scientific activity are nuclear physics, accelerator technology, and elementary particle physics. He was the first to measure the spin and parity of the pi-meson, and he was one of the originators of the method of colliding beams. In 1974 he was President of the American Physics Society. He received the E.O. Lawrence Prize in 1961, the B. Franklin Medal in 1970, as well as other awards.

Vasily Vasilievich Parkhomchuk, doctor of physics and mathematical sciences, is a chief research fellow at the Budker Institute of Nuclear Physics of the Siberian Division of the Russian Academy of Sciences. He graduated from Novosibirsk State University in 1968. His main work is in the area of the physics of charged particle beams. He is one of the originators of the electron cooling method. He has worked at the Institute of Nuclear Physics since 1968.

Stanislav Georgievich Popov is a doctor of physics and mathematical sciences, a professor, and a laboratory head at the Budker Institute of Nuclear Physics of the Siberian Division of the Russian Academy of Sciences. He graduated from the department of physics of Moscow State University in 1959. His main work is in the area of accelerators and experimental nuclear physics. He has worked at the Institute of Nuclear Physics since 1959.

Burton Richter is a professor of physics at Stanford University and, since 1984, the director of the Stanford Linear Accelerator Center. He received the Nobel Prize in physics in 1976. He is a member of the U.S. National Academy of Science. His research interests are high-energy physics and particle accelerators.

Norman Rostoker is a professor of physics at the University of California, Irvine. Previously he was at Cornell University; University of California, San Diego; and General Atomic. He specializes in plasma physics, controlled thermonuclear fusion, and pulsed power physics. He was awarded the Maxwell Prize in Plasma Physics in 1989.

Dmitri Dmitrievich Ryutov, academician, is a chief research fellow and former deputy director (1977-1988) of the Budker Institute of Nuclear Physics of the Siberian Division of the Russian Academy of Sciences. He graduated from the Moscow Institute of Physics and Technology in 1962. From 1962 to 1968 he worked at the Kurchatov Institute of Atomic Energy. He specializes in the area of plasma physics and controlled thermonuclear fusion. He has worked at the Institute of Nuclear Physics since 1968.

Roald Zinnurovich Sagdeev, academician, graduated from Moscow State University in 1955. He worked at the Kurchatov Institute of Atomic Energy until 1961, when he joined the Institute of Nuclear Physics of the Siberian Division of the U.S.S.R. Academy of Sciences. From 1973 to 1988 he was the director of the Space Research Institute (IKI) of the U.S.S.R. Academy of Sciences. At present he is distinguished professor of physics at the University of Maryland, College Park.

Rustam Abel'evich Salimov is a doctor of technical sciences, recipient of the U.S.S.R. State Prize, and a laboratory head at the Budker Institute of Nuclear Physics of the Siberian Division of the Russian Academy of Sciences. He graduated from Novosibirsk State University in 1964. His research is in the area of the physics of charged particle beams and accelerator technology. He has worked at the Institute of Nuclear Physics since 1961.

Roy F. Schwitters is a professor of physics at Harvard University and currently the director of the SSC Laboratory. His research interests are experimental high-energy physics, development of large solid angle detection apparatus for use with high-energy colliding beams, and hadron production in electron-positron collisions.

Veniamin Alexandrovich Sidorov is a corresponding member of the Russian Academy of Sciences, recipient of the Lenin Prize, and deputy director of the Budker Institute of Nuclear Physics of the Siberian Division of the Russian Academy of Sciences. He graduated from Moscow State University in 1953. He specializes in the area of high-energy physics. He was one of the originators of the colliding-beam method. He has worked at the Institute of Nuclear Physics since 1962.

Alexander Nikolaevich Skrinsky is an academician, recipient of the Lenin Prize, academician-secretary of the Department of Nuclear Physics of the Russian Academy of Sciences, and since 1977 the director of the Institute of Nuclear Physics of the Siberian Division of the Russian Academy of Sciences. He graduated from Moscow State University in 1959. His main work is in the areas of elementary particle physics, nuclear physics, the physics of charged particle beams, and accelerator technology. He was one of the originators of the colliding-beam method and the electron cooling method. He has worked at the Institute of Nuclear Physics since 1959.

Karl Strauch is George Vasmer Leverett Professor of Physics Emeritus at Harvard University and, from 1967 to 1974, was the director of the Cambridge Electron Accelerator Laboratory, a joint Harvard-MIT project. His research interests are high-energy reactions and elementary particles.

Lee C. Teng is a senior physicist at Argonne National Laboratory and a member of the board of directors for the Synchrotron Radiation Research Center of Taiwan. He was the director of the Particle Accelerator Division of Argonne National Laboratory from 1962 to 1967 and the head of the Advanced Projects Department of Fermi National Accelerator Laboratory from 1980 to 1987. His research interests are high-energy accelerators and instrumentation.

Emil' Mikhailovich Trakhtenberg is a candidate of technical sciences and a senior scientist. He was head of the Design Department at the Budker Institute of Nuclear Physics of the Siberian Division of the Russian Academy of Sciences until 1992. He graduated from the Moscow Institute of Machine-Tool Construction in 1956. He participated in constructing the VEPP-2, VEPP-3, and VEPP-4 storage rings and was a leading designer for the VEPP-2M and "Siberia" storage rings. He worked at the Institute of Nuclear Physics from 1962.

Tatyana Alexeevna Vsevolozhskaya is a candidate in physics and mathematical sciences and a senior scientist. She graduated from Moscow State University in 1960 and specializes in the physics of charged particle beams and accelerator technology. She has worked at the Budker Institute of Nuclear Physics of the Siberian Division of the Russian Academy of Sciences since 1962.

Boris Grigorievich Yerozolimsky, doctor of physics and mathematical sciences, is a professor and a winner of the U.S.S.R. State Prize. He graduated from the Physics Department of Moscow State University in 1947. From 1948 to 1956 he worked at the Kurchatov Institute of Atomic Energy. From 1957 until 1962 he was a laboratory head at the Institute of Nuclear Physics of the Siberian Division of the U.S.S.R. Academy of Sciences. He participated in the construction of the VEP-1 colliding-beam device. From 1962 to 1982 he was a laboratory head at the Kurchatov Institute of Atomic Energy in Moscow. He was a group leader at the Konstantinov Institute of Nuclear Physics in Leningrad from 1982 to 1991.

Yakov Borisovich Zel'dovich (1914-1987) was an academician, a three-time Hero of Socialist Labor, a Lenin Prize laureate, and a four-time recipient of the U.S.S.R. State Prize. His work was in chemical physics, combustion theory, shock waves and detonation physics, physical chemistry, nuclear and elementary particle physics, astrophysics, and cosmology. He was one of the key scientific contributors to the development of the Soviet nuclear project. He was the founder of a school of relativistic astrophysics.

Curriculum Vitae of G. I. Budker

May 1, 1918

Born in the village of Murafa, Shargorod District, Vinnitsa Region.

1936

Graduated from Secondary School No. 9 in Vinnitsa. Entered the Department of Physics of Moscow State University.

1941

Graduated from Moscow State University with a major in theoretical physics.

1941-1946

Military service in the Soviet Army.

1945

Participant in the All-Union Conference of Army Inventors.

1946

Demobilized from the Soviet Army.

1946-1954

Junior Scientist, and then Senior Scientist, at Laboratory No. 2 of the U.S.S.R. Academy of Sciences (now the I.V. Kurchatov Institute of Atomic Energy), where he worked in the nuclear project.

1947-1949

Instructor in general physics in the Department of Physics and Technology of Moscow State University.

1950

Received the degree of Candidate of Science in Physics and Mathematics.

Awarded the Prize of the U.S.S.R. Council of Ministers for completing special assignments for the government.

1951

Awarded the U.S.S.R. State Prize for his participation in designing the Dubna proton phasotron.

Awarded the Order of Labor Red Banner for completing special assignments for the government.

1952

Proposed the idea of the stabilized electron beam as a new method for charged particle acceleration.

Proposed the idea of plasma confinement in traps with magnetic mirrors ("plugs"), which originated a new approach to the problem of controlled thermonuclear fusion.

1953

Became scientific leader of a group of experimentalists and engineers that was organized at the Institute of Atomic Energy to investigate the possibility of creating a stabilized electron beam. From this group there later grew up the Institute of Nuclear Physics.

1954

Budker's group is transformed into a separate section within the Institute of Atomic Energy.

1955

Received the degree of Doctor of Science in Physics and Mathematics.

1956

Presented a report on the stabilized electron beam at the International Conference in Geneva.

Started developing the method of colliding beams.

1956-1958

Professor in the Department of Theoretical Physics at the Moscow Institute of Engineering Physics.

1958-1977

Director of the Institute of Nuclear Physics, which he had founded.

Actively participated in organizing the Siberian Division of the U.S.S.R. Academy of Sciences and also Novosibirsk State University.

1958

Elected as Corresponding Member of the U.S.S.R. Academy of Sciences.

1958-1975

Member of the Presidium of the Siberian Division of the U.S.S.R. Academy of Sciences.

1959

Organized and headed the Department of General Physics at Novosibirsk State University.

1960

Soviet delegate to the International (Rochester) Conference on High-Energy Physics.

1962

Chairman of the jury for the First All-Siberian Physics-Mathematics Olympiad for secondary school students. One of those who initiated setting up a special physics and mathematics school within Novosibirsk State University.

1962-1977

Head of the Department of Nuclear Physics, which he organized at Novosibirsk State University.

1963

Participant at the International Conference on Accelerators in Dubna. The first report about the work on colliding beams at the Institute of Nuclear Physics. The beginning of collaboration with scientists from accelerator centers in Europe and the U.S.

Began developing industrial accelerators for the national economy.

1963-1977

Member of the Executive Committee of the Nuclear Physics Division of the U.S.S.R. Academy of Sciences.

1964

Elected as Academician.

1965

> The first experiments to check quantum electrodynamics at small distances, using electron-electron scattering in the VEP-1 device.
>
> Participant at the International Conference on Accelerators in Italy.

1966

> Proposed and developed the simple and elegant idea of electron cooling of beams of heavy particles.
>
> Leader of the Soviet delegation to the International Conference on Colliding-Beam Accelerators. The first report about electron cooling.
>
> Visited a number of leading laboratories in the United States at the invitation of the U.S. National Academy of Science.

1967

> First experiments on the VEPP-2 electron-positron colliding-beam device.
>
> Awarded the Lenin Prize for developing the method of colliding beams.
>
> Awarded the Order of Lenin for successful achievements in developing Soviet science and applying scientific accomplishments to the national economy.
>
> Visited the European Center for Nuclear Research (CERN) at the invitation of its directorate.

1968

Presented a speech at the International Conference on Plasma Physics and Controlled Nuclear Fusion Research in Novosibirsk.

1969

Visited a number of scientific centers in Japan at the invitation of the Japanese Academy of Science.

1972

Work using synchrotron radiation is begun at the Institute of Nuclear Physics.

1973

Presented an invited report at the Fourth European Conference on Plasma Physics and Controlled Fusion, reviewing work at the Institute of Nuclear Physics on multiple-mirror plasma devices.

1974

Experiments on the VEPP-2M electron-positron storage ring, which has the world's highest luminosity in its energy range, are begun.

Experimental confirmation of the method of electron cooling of a proton beam.

1975

Awarded the Order of the October Revolution for achievements in the development of Soviet science.

May, 1977

> An electron beam is circulated in VEPP-4, the world's largest storage ring.

July 4, 1977

> Died in Novosibirsk.

List of Publications of G.I. Budker

1941

1. G.I. Budker, "Energy-momentum tensor in a medium," Master's degree thesis, Moscow State University (1941) (in Russian).

1946-1954

[During 1946-1954, he wrote 46 internal reports, some of which were later published.]

1956

2. G.I. Budker, "Relativistic stabilized electron beam. I: Physical principles and theory," in *Proc. CERN Symposium on High-Energy Accelerators and Pion Physics, Geneva, June, 1956* (CERN, Geneva, 1956), Vol. I, pp. 68-75. Also published in *Atomnaya Energiya* **1** (5), 9-19 (1956) (in Russian). Also published in *Collective Methods of Acceleration: Papers Presented at the 3rd International Conference on Collective Methods of Acceleration Dedicated to G.I. Budker, Irvine, 1978* (Harwood Academic Publ., New York, 1979), pp. 1-22.

3. G.I. Budker and A.A. Naumov, "Relativistic stabilized electron beam. II: Brief review of experimental work," in *Proc. CERN Symposium on High-Energy Accelerators and Pion Physics, Geneva, June, 1956* (CERN, Geneva, 1956), Vol. I, pp. 76-79.

4. S.T. Belyaev and G.I. Budker, "The relativistic kinetic equation," *Doklady Akademii Nauk SSSR* **107** (6), 807-810 (1956) [*Soviet Physics–Doklady* **1** (2), 218-222 (1956)].

1958

5. G.I. Budker, "Questions associated with the drift of particles in a toroidal magnetic thermonuclear reactor" (work completed in 1951), first published in *Fizika Plazmy i Problemy Upravlyayemykh Termoyadernykh Reaktsii* (U.S.S.R. Academy of Science, Moscow and Leningrad, 1958), Vol. I, pp. 66-76 [*Plasma Physics and the Problem of Controlled Thermonuclear Reactions*, edited by M.A. Leontovich (Pergamon Press, London, 1961), Vol. I, pp. 78-88].

6. G.I. Budker, "The betatron method of heating plasma to high temperatures" (work completed in 1951), first published in *Fizika Plazmy i Problemy Upravlyayemykh Termoyadernykh Reaktsii* (U.S.S.R. Academy of Science, Moscow and Leningrad, 1958), Vol. I, pp. 122-129 [*Plasma Physics and the Problem of Controlled Thermonuclear Reactions*, edited by M.A. Leontovich (Pergamon Press, London, 1961), Vol. I, pp. 145-152].

7. S.I. Braginsky and G.I. Budker, "Physical phenomena in the process of ignition of a discharge with incomplete ionization" (work completed in 1952), first published in *Fizika Plazmy i Problemy*

Upravlyayemykh Termoyadernykh Reaktsii (U.S.S.R. Academy of Science, Moscow and Leningrad, 1958), Vol. 1, pp. 186-206 [*Plasma Physics and the Problem of Controlled Thermonuclear Reactions*, edited by M.A. Leontovich (Pergamon Press, London, 1961), Vol. I, pp. 225-254].

8. G.I. Budker, "Electrical breakdown in a gas in the presence of strong external time-variable magnetic field" (work completed in 1952), first published in *Fizika Plazmy i Problemy Upravlyayemykh Termoyadernykh Reaktsii* (U.S.S.R. Academy of Science, Moscow and Leningrad, 1958), Vol. I, pp. 214-221 [*Plasma Physics and the Problem of Controlled Thermonuclear Reactions*, edited by M.A. Leontovich (Pergamon Press, London, 1961), Vol. I, pp. 263-272].

9. G.I. Budker, "Thermonuclear reactions in a potential well of a negative charge" (work completed in 1952), first published in *Fizika Plazmy i Problemy Upravlyayemykh Termoyadernykh Reaktsii* (U.S.S.R. Academy of Science, Moscow and Leningrad, 1958), Vol. I, pp. 243-248 [*Plasma Physics and the Problem of Controlled Thermonuclear Reactions*, edited by M.A. Leontovich (Pergamon Press, London, 1961), Vol. I, pp. 295-301].

10. S.T. Belyaev and G.I. Budker, "Relativistic plasma in variable fields" (work completed in 1953), first published in *Fizika Plazmy i Problemy Upravlyayemykh Termoyadernykh Reaktsii* (U.S.S.R. Academy of Science, Moscow and Leningrad, 1958), Vol. II, pp. 283-329 [*Plasma Physics and the Problem of Controlled Thermonuclear Reactions*, edited by M.A. Leontovich (Pergamon Press, London, 1959), Vol. II, pp. 383-430].

11. S.T. Belyaev and G.I. Budker, "Boltzmann's equation for electron gas in which collisions are infrequent" (work completed in 1954), first published in *Fizika Plazmy i Problemy Upravlyayemykh Termoyadernykh Reaktsii* (U.S.S.R. Academy of Science, Moscow and Leningrad, 1958), Vol. II, pp. 330-354 [*Plasma Physics and the Problem of Controlled Thermonuclear Reactions*, edited by M.A. Leontovich (Pergamon Press, London, 1959), Vol. II, pp. 431-457].

12. G.I. Budker, "Thermonuclear reactions in a system with magnetic stoppers and the problem of direct transformation of thermonuclear energy into electrical energy" (work completed in 1954), first published in *Fizika Plazmy i Problemy Upravlyayemykh Termoyadernykh Reaktsii* (U.S.S.R. Academy of Science, Moscow and

Leningrad, 1958), Vol. III, pp. 3-31 [*Plasma Physics and the Problem of Controlled Thermonuclear Reactions*, edited by M.A. Leontovich (Pergamon Press, London, 1959), Vol. III, pp. 1-33].

13. G.I. Budker, "Some problems of the spatial stability of a ring current in a plasma" (work completed in 1951), first published in *Fizika Plazmy i Problemy Upravlyayemykh Termoyadernykh Reaktsii* (U.S.S.R. Academy of Science, Moscow and Leningrad, 1958), Vol. III, pp. 32-40 [*Plasma Physics and the Problem of Controlled Thermonuclear Reactions*, edited by M.A. Leontovich (Pergamon Press, London, 1959), Vol. III, pp. 34-44].

14. S.T. Belyaev and G.I. Budker, "Multi-quantum recombination in an ionized gas" (work completed in 1955), first published in *Fizika Plazmy i Problemy Upravlyayemykh Termoyadernykh Reaktsii* (U.S.S.R. Academy of Science, Moscow and Leningrad, 1958), Vol. III, pp. 41-49 [*Plasma Physics and the Problem of Controlled Thermonuclear Reactions*, edited by M.A. Leontovich (Pergamon Press, London, 1959), Vol. III, pp. 45-55].

1962

15. G.I. Budker, B.G. Yerozolimsky, and A.A. Naumov, "The VEP-1 colliding-beam facility and experiments on elastic scattering of electrons," in *Proceedings of the Conference on Physics and Technology of the Colliding-Beam Method, Kharkov, 1962* (Kharkov, 1962), Part 2, pp. 5-32 (in Russian).

16. G.I. Budker, A.Kh. Kadymov, A.A. Naumov, V.S. Panasyuk, S.G. Popov, and A.N. Skrinsky, "Ironless pulsed synchrotron for energies of 70-100 MeV," in *Proceedings of the Conference on Physics and Technology of the Colliding-Beam Method, Kharkov, 1962* (Kharkov, 1962), Part 2, pp. 91-110 (in Russian).

17. G.I. Budker, P.I. Medvedev, Yu.A. Mostovoy, I.M. Samoilov, and A.A. Sokolov, "High-current ironless single-turn synchrotron of 400 MeV," in *Proceedings of the Conference on Physics and Technology of the Colliding-Beam Method, Kharkov, 1962* (Kharkov, 1962), Part 2, pp. 111-134 (in Russian).

1963

18. G.I. Budker, "On the way to antimatter," in *Izvestiya* newspaper, August 23, 1963 (August 22 Moscow evening issue) (in Russian).

1964

19. E.A. Abramyan, V.L. Auslender, V.N. Bayer, G.A. Blinov, L.N. Bondarenko, G.I. Budker, S.B. Vasserman, V.V. Vecheslavov, G.I. Dimov, B.G. Yerozolimsky, A.V. Kiselyev, L.S. Korobeinikov, E.A. Kushnirenko, A.A. Livshitz, E.S. Mironov, A.A. Naumov, A.P. Onuchin, V.S. Panasyuk, V.A. Papadichev, S.G. Popov, I.Ya. Protopopov, S.N. Rodionov, V.A. Sidorov, G.I. Sil'vestrov, A.N. Skrinsky, V.S. Synakh, A.G. Khabakhpashev, and L.I. Yudin, "Work on colliding electron-electron, positron-electron, and proton-proton beams at the Institute of Nuclear Physics of the Siberian Division of the U.S.S.R. Academy of Sciences," in *Proceedings of the International Conference on Accelerators, Dubna, 1963* (Atomizdat, Moscow, 1964), pp. 274-287 (in Russian).
20. G.I. Budker and G.I. Dimov, "Charge exchange injection of protons into circular accelerators," in *Proceedings of the International Conference on Accelerators, Dubna, 1963* (Atomizdat, Moscow, 1964), pp. 993-996 (in Russian).
21. E.A. Abramyan, I.E. Bender, L.N. Bondarenko, G.I. Budker, G.B. Glagolev, A. Kh. Kadymov, N.G. Kon'kov, I.N. Meshkov, Yu.A. Mostovoy, A.A. Naumov, O.A. Nezhevenko, G.N. Ostreiko, V.E. Pal'chikov, V.S. Panasyuk, V.V. Petrov, S.G. Popov, I.Ya. Protopopov, S.N. Rodionov, I.M. Samoilov, A.N. Skrinsky, A.A. Sokolov, I.Ya. Timoshin, and L.I. Yudin, "Work on high-current accelerators at the Institute of Nuclear Physics of the Siberian Division of the U.S.S.R. Academy of Sciences," in *Proceedings of the International Conference on Accelerators, Dubna, 1963* (Atomizdat, Moscow, 1964), pp. 1065-1072 (in Russian).
22. G.I. Budker, "Creation of accelerators with colliding beams," *Vestnik Akademii Nauk SSSR*, No. 6, pp. 31-36 (1964) (in Russian).
23. G.I. Budker, "Taming plasma," in *Izvestiya* newspaper, Sept. 6, 1964 (Sept. 5 Moscow evening issue) (in Russian).

1965

24. E.A. Abramyan, G.I. Budker, G.B. Glagolev, and A.A. Naumov, "A betatron with spiral electron storage," *JTP* **35** (4), pp. 605-611 (1965) [*Soviet Physics–Technical Physics* **10** (4), 477-481 (1965)].

25. E.A. Abramyan, I.E. Bender, G.I. Budker, A.Kh. Kadymov, A.A. Naumov, and V.S. Panasyuk, "Pulsed iron-free synchrotron," *JTP* **35** (4), 612-617 (1965) [*Soviet Physics–Technical Physics* **10** (4), 482-485 (1965)].

26. V.L. Auslender, G.A. Blinov, G.I. Budker, G.I. Dimov, M.M. Karliner, A.V. Kiselyov, E.A. Kushnirenko, A.A. Livshitz, S.I. Mishnev, A.A. Naumov, A.P. Onuchin, V.S. Panasyuk, Yu.N. Pestov, S.G. Popov, Yu.K. Sviridov, V.A. Sidorov, G.I. Sil'vestrov, A.N. Skrinsky, B.N. Sukhina, I.Ya. Timoshin, G.M. Tumaikin, A.G. Khabakhpashev, and I.A. Shekhtman, "The development of accelerators at Novosibirsk," *Atomnaya Energiya* **19** (6), 497-510 (1965) [*Soviet Atomic Energy* **19** (6), 1465-1481 (1965)]. Also published in *Proc. 5th Intern. Conf. on High-Energy Accelerators, Frascati, 1965* (Roma, 1966), pp. 389-402.

27. G.I. Budker, "Radiation for industry," in *Izvestiya* newspaper, August 1, 1965 (July 31 Moscow evening issue) (in Russian).

1966

28. G.I. Budker, P.I. Medvedev, Yu.A. Mostovoy, O.A. Nezhevenko, A.B. Nelidov, G.N. Ostreiko, V.S. Panasyuk, I.M. Samoilov, and A.A. Sokolov, "The BSB iron-free single-turn injection synchrotron," *JTP* **36** (9), 1523-1535 (1966) [*Soviet Physics–Technical Physics* **11** (9), 1139-1148 (1967)].

29. G.I. Budker, V.I. Volosov, S.S. Moiseev, V.E. Pal'chikov, and F.A. Tsel'nik, "Plasma with relativistic electrons in a magnetic mirror trap," in *Plasma Physics and Controlled Nuclear Fusion Research (Proc. Conf. on Plasma Physics and Controlled Nuclear Fusion Research, Culham, 1965)* (International Atomic Energy Agency, Vienna, 1966), Vol. II, pp. 245-258.

30. S.G. Alikhanov, G.I. Budker, G.N. Kichigin, and A.V. Komin, "Metal shell collapse in a magnetic field," *Zhurnal Prikladnoi*

Mekhaniki i Teoreticheskoi Fiziki **7** (4), 38-41 (1966) [*Journal of Applied Mechanics and Technical Physics*, **7** (4), 24-26 (1966)].

31. G.I. Budker, "Accelerators with colliding particle beams," *Vestnik Akademii Nauk SSSR*. No. 4, 30-43 (1966) (in Russian).

32. G.I. Budker, "Accelerators with colliding particle beams," *Uspekhi Fizicheskikh Nauk* **89** (4), 533-547 (1966) [*Soviet Physics–Uspekhi* **9** (4), 534-542 (1967)]. Presented at the general meeting of the U.S.S.R. Academy of Sciences, February 8, 1966.

33. G.I. Budker, "Experiments with antimatter," in *Izvestiya* newspaper, Oct. 18, 1966 (Oct. 17 Moscow evening issue) (in Russian).

34. S.G. Alikhanov, G.I. Budker, A.V. Komin, V.A. Polyakov, and B.S. Estrin, "Experiments with dense hot plasma," in *Proc. 7th Intern. Conf. on Phenomena in Ionized Gases, Beograd, 1965* (Gradevinska Kniga, Beograd, 1966), Vol. I, pp. 776-780.

35. V.L. Auslender, G.I. Budker, M.M. Karliner, A.A. Naumov, S.G. Popov, A.N. Skrinsky, and I.A. Shekhtman, "Phase instability of the intense electron beam in a storage ring," *Atomnaya Energiya* **20** (3), 210-213 (1966) [*Soviet Atomic Energy* **20** (3), 240-243 (1966)].

36. G.I. Budker, E.A. Kushnirenko, A.A. Naumov, A.P. Onuchin, S.G. Popov, V.A. Sidorov, A.N. Skrinsky, and G.M. Tumaikin, "Electron-electron scattering at 2 × 135 MeV," in *Proc. Symposium International sur les Anneaux de Collisions a Electrons et Positrons, Saclay, 1966* (Paris, 1966), pp. V/2/1-V/2/10.

37. G.I. Budker, A.V. Kiselev, N.G. Kon'kov, A.A. Naumov, V.I. Nifontov, G.N. Ostreiko, V.S. Panasyuk, V.V. Petrov, L.I. Yudin, and G.I. Yasnov, "Starting up the synchrotron B-3M injector for positron-electron storage ring," in *Proc. 5th International Conference on High-Energy Accelerators, Frascati, 1965* (Roma, 1966), pp. 455-458.

38. G.I. Budker, A.V. Kiselev, N.G. Kon'kov, A.A. Naumov, V.I. Nifontov, G.N. Ostreiko, V.S. Panasyuk, V.V. Petrov, L.I. Yudin, and G.S. Yasnov, "Starting up the B-3M synchrotron injector for the positron-electron storage ring," *Atomnaya Energiya* **20** (3), 206-210 (1966) [*Soviet Atomic Energy* **20** (3), 235-239 (1966)].

39. G.I. Budker, "Status report of work on storage rings at Novosibirsk," in *Proc. Symposium International sur les Anneaux de Collisions a Electrons et Positrons, Saclay, 1966* (Paris, 1966), pp. II/1/1-II/1/5.

1967

40. G.I. Budker, "State of the art for colliding beams at the Nuclear Physics Institute of the Siberian Section of the Academy of Sciences of the U.S.S.R.," *Atomnaya Energiya* **22** (3), 163-164 (1967) [*Soviet Atomic Energy* **22** (3), 198-199 (1967)].

41. G.I. Budker, E.A. Kushnirenko, A.A. Naumov, A.P. Onuchin, S.G. Popov, V.A. Sidorov, A.N. Skrinsky, and G.M. Tumaikin, "Electron-electron scattering at 2×135 MeV," *Atomnaya Energiya* **22** (3), 164-168 (1967) [*Soviet Atomic Energy* **22** (3), 200-204 (1967)].

42. V.L. Auslender, G.I. Budker, A.A. Naumov, Yu.N. Pestov, V.A. Sidorov, A.N. Skrinsky, and A.G. Khabakhpashev, "First experiments on the VEPP-2 positron-electron storage system," *Atomnaya Energiya* **22** (3), 173-175 (1967) [*Soviet Atomic Energy* **22** (3), 210-212 (1967)]. Also published in *Proc. Symposium International sur les Anneaux de Collisions a Electrons et Positrons, Saclay, 1966* (Paris, 1966), pp. V/4/1-V/4/7.

43. G.I. Budker, "An effective method of damping particle oscillations in proton and antiproton storage rings," *Atomnaya Energiya* **22** (5), 346-348 (1967) [*Soviet Atomic Energy* **22** (5), 438-440 (1967)].

44. G.I. Budker, G.I. Dimov, and V.G. Dudnikov, "Experiments on producing intensive proton beams by means of the method of charge exchange injection," *Atomnaya Energiya* **22** (5), 348-355 (1967) [*Soviet Atomic Energy* **22** (5), 441-448 (1967)]. Also published in *Proc. Symposium International sur les Anneaux de Collisions a Electrons et Positrons, Saclay, 1966* (Paris, 1966), pp. VIII/6/1-VIII/6/ 12.

45. S.G. Alikhanov, V.G. Belan, G.I. Budker, A.I. Ivanchenko, and G.N. Kichigin, "Creating megagauss fields by the method of magnetodynamic accumulation," *Atomnaya Energiya* **23** (6), 536-541 (1967) [*Soviet Atomic Energy* **23** (6), 1307-1311 (1967)].

46. G.I. Budker, A.P. Onuchin, S.G. Popov, and G.M. Tumaikin, "Experiments with the target in electron storage ring," *Yadernaya Fizika* **6** (4), 775-779 (1967) [*Sov. Journ. Nucl. Phys.* **6** (4), 563-566 (1968)].

47. G.I. Budker, E.A. Kushnirenko, R.L. Lebedev, A.A. Naumov, A.P. Onuchin, S.G. Popov, V.A. Sidorov, A.N. Skrinsky, and G.M. Tumaikin, "Checking quantum electrodynamics in electron-electron scattering," *Yadernaya Fizika* **6** (6), 1221-1225 (1967) [*Sov. Journ. Nucl. Phys.* **6** (6), 889-892 (1968)].

48. V.L. Auslender, G.I. Budker, Yu.N. Pestov, V.A. Sidorov, A.N. Skrinsky, and A.G. Khabakhpashev, "Investigation of the ρ-meson resonance with electron-positron colliding beams," *Physics Letters* **25B** (6), 433-435 (1967).

49. G.I. Budker, I.Ya. Protopopov, and A.N. Skrinsky, "3.5 BeV electron-positron colliding-beam machine VEPP-3 at Novosibirsk," in *Proc. 6th Intern. Conf. on High-Energy Accelerators, Cambridge, 1967*, pp. 102-103.

1968

50. G.I. Budker, "Keys to the Sun," in *Trud* (Labor) newspaper, Aug. 8, 1968 (in Russian).

1969

51. G.I. Budker, "The concluding speech at the 3rd International Conference on Plasma Physics and Controlled Nuclear Fusion Research," in *Plasma Physics and Controlled Nuclear Fusion Research (Proc. 3rd Intern. Conf. on Plasma Physics and Controlled Nuclear Fusion Research, Novosibirsk, 1968)* (International Atomic Energy Agency, Vienna, 1969), Vol. I, pp. 41-42 (in Russian), pp. 43-44 (in English).

52. V.L. Auslender, G.I. Budker, E.V. Pakhtusova, Yu.N. Pestov, V.A. Sidorov, A.N. Skrinsky, and A.G. Khabakhpashev, "Study of the ρ-meson resonance in colliding electron-positron beams," *Yadernaya Fizika* **9** (1), 114-119 (1969) [*Sov. Journ. Nucl. Phys.* **9** (1), 69-72 (1969)].

53. V.E. Balakin, G.I. Budker, Yu.V. Korshunov, S.I. Mishnev, E.V. Pakhtusova, Yu.N. Pestov, V.A. Sidorov, A.N. Skrinsky, G.M. Tumaikin, and A.G. Khabakhpashev, "Preliminary results of studies of φ-meson resonance using electron-positron colliding beams: Vector mesons and electromagnetic interactions," in *Proc. of International Seminar, Dubna, 1969* (Dubna, 1969), pp. 479-488 (in Russian).

54. G.I. Budker, "Science expands horizons: microworld economics," in *Pravda* newspaper, Feb. 27, 1969 (in Russian).

55. G.I. Budker, "Buy an accelerator!" in *Ogonyok*, No. 19, p. 17 (1969) (in Russian).

1970

56. G.I. Budker, N.A. Kuznetsov, B.V. Levichev, I.Ya. Protopopov, and A.N. Skrinsky, "Magnetic system of the VEPP-3 storage ring," in *Proc. All-Union Conf. on Charged Particle Accelerators, Moscow, 1968* (VINITI, Moscow, 1970), Vol. I, pp. 270-273 (in Russian).

57. G.I. Budker, T.A. Vsevolozhskaya, and G.I. Sil'vestrov, "Monochromatization of a positron beam at a wedge-shaped target for increasing the trapping coefficient in a storage ring," in *Proc. of All-Union Conference on Charged Particle Accelerators, Moscow, 1968* (VINITI, Moscow, 1970), Vol. I, pp. 544-547 (in Russian).

58. A.I. Arenshtam, G.I. Budker, I.N. Meshkov, V.G. Ponomarenko, and A.N. Skrinsky, "A system for electron cooling," in *Proc. of All-Union Conf. on Charged Particle Accelerators, Moscow, 1968* (VINITI, Moscow, 1970), Vol. II, pp. 400-403 (in Russian).

59. G.I. Budker, "Studies on colliding beams at the Siberian Institute of Nuclear Physics: Status of work and prospects," in *Proc. of All-Union Conf. on Charged Particle Accelerators, Moscow, 1968* (VINITI, Moscow, 1970), Vol. II, pp. 1017-1023 (in Russian).

60. G.I. Budker, "Accelerators and colliding beams," in *Proc. 7th Intern. Conf. on High-Energy Charged Particle Accelerators, Yerevan, 1969* (Armenian Academy of Science, Yerevan, 1970), Vol. I, pp. 33-39 (in Russian).

61. V.L. Auslender, G.I. Budker, I.B. Vasserman, N.S. Dikansky, M.M. Karliner, M.D. Malev, S.I. Mishnev, V.A. Sidorov, A.N. Skrinsky, G.M. Tumaikin, A.G. Khabakhpashev, Yu.M. Shatunov, and I.A.

Shekhtman, "Reconstruction of the VEPP-2 device with colliding electron-positron beams," in *Proc. 7th Intern. Conf. on High-Energy Charged Particle Accelerators, Yerevan, 1969* (Armenian Academy of Science, Yerevan, 1970), Vol. II, pp. 26-36 (in Russian).

62. G.I. Budker, I.Ya. Protopopov, and A.N. Skrinsky, "A device with colliding electron-positron beams for an energy of 3.5 GeV (VEPP-3)," in *Proc. 7th Intern. Conf. on High-Energy Charged Particle Accelerators, Yerevan, 1969* (Armenian Academy of Science, Yerevan, 1970), Vol. II, pp. 37-47 (in Russian).

63. G.I. Budker, "By way of antiplasma to antimatter," in *Za Nauku v Sibiri* (Pro Siberian Science) newspaper, January 14, 1970 (in Russian).

64. V.E. Balakin, G.I. Budker, I.B. Vasserman, O.S. Koifman, L.M. Kurdadze, S.I. Mishnev, A.P. Onuchin, S.I. Serednyakov, V.A. Sidorov, A.N. Skrinsky, G.M. Tumaikin, V.F. Turkin, A.G. Khabakhpashev, and Yu.M. Shatunov, "Experiments on colliding positron-electron beams at the energies $2 \times 590, 2 \times 630, 2 \times 670$ MeV," Institute of Nuclear Physics of the Siberian Division of the U.S.S.R. Academy of Sciences, Report No. 62-70 (Novosibirsk, 1970) (in Russian). Presented at the 15th International Conference on High-Energy Physics, Kiev, 1970.

65. S.T. Belyaev, G.I. Budker, and S.G. Popov, "The possibility of using storage rings with internal thin targets," in *High-Energy Physics and Nuclear Structure (Proceedings of the 3rd International Conference on High-Energy Physics and Nuclear Structure, New York, 1969)* (Plenum Press, New York, 1970), pp. 606-609.

1971

66. G.I. Budker, "Antimatter: dreams and reality," in *Izvestiya* newspaper, January 16, 1971 (Jan. 15 Moscow evening issue) (in Russian).

67. G.I. Budker, V.V. Mirnov, and D.D. Ryutov, "The influence of corrugation of the magnetic field on the expansion and cooling of a dense plasma," *Pis'ma v Zhurnal Eksperimental'noi i Teoreticheskoi Fiziki* **14** (5), 320-322 (1971) [*Soviet Physics–JETP Letters* **14** (5), 212-215 (1971)].

68. V.E. Balakin, G.I. Budker, E.V. Pakhtusova, V.A. Sidorov, A.N. Skrinsky, G.M. Tumaikin, and A.G. Khabakhpashev, "Experiment on 2γ-quantum annihilation on the VEPP-2," *Physics Letters* **34B** (1), 99-100 (1971).

69. V.E. Balakin, G.I. Budker, E.V. Pakhtusova, V.A. Sidorov, A.N. Skrinsky, G.M. Tumaikin, and A.G. Khabakhpashev, "Investigation of the φ-meson resonance at electron-positron colliding beams," *Physics Letters* **34B** (4), 328-332 (1971). Presented at the XV Intern. Conf. on High-Energy Physics, Kiev, 1970.

70. V.E. Balakin, G.I. Budker, L.M. Kurdadze, A.P. Onuchin, E.V. Pakhtusova, S.I. Serednyakov, V.A. Sidorov, A.N. Skrinsky, and A.G. Khabakhpashev, "Test of quantum electrodynamics by $e^+e^- \to \mu^+\mu^-$," *Physics Letters* **37B** (4), 435-437 (1971). Presented at the XV Intern. Conf. on High-Energy Physics, Kiev, 1970.

1972

71. G.I. Budker, V.V. Mirnov, and D.D. Ryutov, "Gasdynamics of a dense plasma in a multiple-mirror magnetic field," in *Proc. Intern. Conf. on Plasma Theory, Kiev, 1971* (Institute of Theoretical Physics of Academy of Science of Ukraine, Kiev, 1972), pp. 145-151 (in Russian).

72. G.I. Budker, V.I. Kudelainen, I.N. Meshkov, V.G. Ponomarenko, S.G. Popov, R.A. Salimov, A.N. Skrinsky, and B.M. Smirnov, "Electron beam for experiments on electron cooling," in *Proc. 2nd All-Union Conf. on Charged Particle Accelerators, Moscow, 1970* (Nauka, Moscow, 1972), Vol. I, pp. 31-33 (in Russian).

73. G.I. Budker, T.A. Vsevolozhskaya, G.I. Sil'vestrov, and A.N. Skrinsky, "A scheme for obtaining antiprotons for a device with colliding proton-antiproton beams," in *Proc. 2nd All-Union Conf. on Charged Particle Accelerators, Moscow, 1970* (Nauka, Moscow, 1972), Vol. II, pp. 196-198 (in Russian).

74. V.E. Balakin, G.I. Budker, L.M. Kurdadze, A.P. Onuchin, E.V. Pakhtusova, S.I. Serednyakov, V.A. Sidorov, and A.N. Skrinsky, "Measurement of the electron-positron annihilation cross-section into $\pi^+\pi^-$, K^+K^--pairs at the total energy 1.18–1.34 GeV," *Physics Letters* **41B** (2), 205-208 (1972).

75. G.I. Budker, "Good student–good scientist," in *Izvestiya* newspaper, February 10, 1972 (in Russian).

1973

76. V.E. Balakin, L.M. Barkov, V.M. Borovikov, G.I. Budker, I.B. Vasserman, E.I. Zinin, M.M. Karliner, I.A. Koop, A.A. Livshitz, A.P. Lysenko, S.I. Mishnev, V.A. Sidorov, A.N. Skrinsky, E.M. Trakhtenberg, G.M. Tumaikin, and Yu.M. Shatunov, "The new device with colliding electron-positron beams at the Novosibirsk Institute of Nuclear Physics (VEPP-2M)," in *Proc. 3rd All-Union Conf. on Charged Particle Accelerators, Moscow, 1972* (Nauka, Moscow, 1973), Vol. I, pp. 318-323 (in Russian).

77. G.I. Budker, V.V. Danilov, E.P. Kruglyakov, D.D. Ryutov, and E.V. Shun'ko, "Experiments on plasma confinement in a magnetic multimirror trap," *Zhurnal Eksperimental'noi i Teoreticheskoi Fiziki* **65** (2), 562-574 (1973) [*Soviet Physics–JETP* **38** (2), 276-282 (1974)].

78. G.I. Budker, V.V. Danilov, E.P. Kruglyakov, D.D. Ryutov, and E.V. Shun'ko, "Experiments on the containment of an alkali plasma in a corrugated magnetic field," *Pis'ma v Zhurnal Eksperimental'noi i Teoreticheskoi Fiziki* **17** (2), 117-120 (1973) [*Soviet Physics–JETP Letters* **17** (2), 81-84 (1973)].

1974

79. G.I. Budker, E.P. Kruglyakov, V.V. Mirnov, and D.D. Ryutov, "On the possibility of a thermonuclear reactor with dense plasma confined by a corrugated magnetic field," in *System Analysis and Construction of Thermonuclear Stations: Reports of Soviet Specialists at the U.S.S.R.-U.S.A. Joint Seminar, 1974* (Leningrad, 1974), No. 11, 12 pages (in Russian).

80. G.I. Budker, "Academician G.I. Budker speaks about the importance of a scientific school," in *Vozrast Poznaniya* (The Age of Knowledge) (Moscow, 1974), pp. 124-142 (in Russian).

81. G.I. Budker, "Controlled thermonuclear fusion in installations with a dense plasma," *Priroda* (Nature), No. 5, 14-21 (1974) (in Russian).
82. G.I. Budker, "Thermonuclear fusion in installations with a dense plasma," in *Proc. 6th Europ. Conf. on Controlled Fusion and Plasma Physics, Moscow, 1973* (Moscow, 1974), Vol. II, pp. 136-153.
83. G.I. Budker, "Expected Unexpectedness," in *Nedelya*, a weekly supplement to *Izvestiya* newspaper, Issue No. 1 (1974) (in Russian).

1975

84. G.I. Budker, Ya.S. Derbenev, N.S. Dikansky, V.V. Parkhomchuk, D.V. Pestrikov, and A.N. Skrinsky, "Kinetics of electron cooling," in *Proc. 4th All-Union Conf. on Charged Particle Accelerators, Moscow, 1974* (Nauka, Moscow, 1975), Vol. II, pp. 300-303 (in Russian).
85. V.V. Anashin, G.I. Budker, N.S. Dikansky, V.I. Kudelainen, I.N. Meshkov, V.V. Parkhomchuk, D.V. Pestrikov, V.G. Ponomarenko, R.A. Salimov, A.N. Skrinsky, and B.N. Sukhina, "A device for electron cooling experiments," in *Proc. 4th All-Union Conf. on Charged Particle Accelerators, Moscow, 1974* (Nauka, Moscow, 1975), Vol. II, pp. 304-308 (in Russian).
86. G.I. Budker, N.S. Dikansky, V.I. Kudelainen, I.N. Meshkov, V.V. Parkhomchuk, D.V. Pestrikov, A.N. Skrinsky, and B.N. Sukhina, "First experiments on electron cooling," in *Proc. 4th All-Union Conf. on Charged Particle Accelerators, Moscow, 1974* (Nauka, Moscow, 1975), Vol. II, pp. 309-312 (in Russian).
87. G.I. Budker, I.B. Vasserman, V.G. Veshcherevich, A.N. Kirpotin, I.A. Koop, L.M. Kurdadze, A.P. Lysenko, S.I. Mishnev, V.N. Osipov, V.M. Petrov, A.A. Polunin, I.K. Sedlyarov, V.A. Sidorov, S.I. Serednyakov, A.N. Skrinsky, E.M. Trakhtenberg, G.M. Tumaikin, V.F. Turkin, Yu.M. Shatunov, I.A. Shekhtman, and A.G. Khabakhpashev, "The status of work on the VEPP-2M device with colliding electron-positron beams," in *Proc. 4th All-Union Conf. on Charged Particle Accelerators, Moscow, 1974* (Nauka, Moscow, 1975), Vol. II, pp. 313-317 (in Russian).

88. V.M. Aul'chenko, G.I. Budker, I.B. Vasserman, I.A. Koop, L.M. Kurdadze, V.P. Kutov, A.P. Lysenko, S.I. Mishnev, E.V. Pakhtusova, S.I. Serednyakov, V.A. Sidorov, A.N. Skrinsky, G.M. Tumaikin, A.G. Khabakhpashev, A.G. Chilingarov, Yu.M. Shatunov, B.A. Shvartz, and S.I. Eidelman, "Initial experiments on the VEPP-2M electron-positron storage ring," in *Proc. 5th Intern. Symp. on High-Energy and Elementary Particle Physics, Warsaw, 1975* (Dubna, 1975), pp. 163-171.

89. G.I. Budker, V.V. Danilov, V.A. Kornilov, E.P. Kruglyakov, V.N. Lukyanov, V.V. Mirnov, and D.D. Ryutov, "Plasma confinement in multimirror magnetic field," in *Plasma Physics and Controlled Nuclear Fusion Research (Proc. 5th Intern. Conf. on Plasma Physics and Controlled Nuclear Fusion Research, Tokyo, 1974)* (International Atomic Energy Agency, Vienna, 1975), Vol. II, pp. 763-776 [*Nucl. Fus. Suppl.*, pp. 147-153 (1975)].

90. G.I. Budker, E.P. Kruglyakov, V.V. Mirnov, and D.D. Ryutov, "On the possibility of a thermonuclear reactor with dense plasma confined by a corrugated magnetic field," *Izvestiya AN SSSR: Series on "Energetics and Transport,"* No. 6, pp. 35-40 (1975) (in Russian).

91. G.I. Budker, Ya.S. Derbenev, N.S. Dikansky, V.I. Kudelainen, I.N. Meshkov, V.V. Parkhomchuk, D.V. Pestrikov, B.N. Sukhina, and A.N. Skrinsky, "Experiments on electron cooling," *IEEE Trans. Nucl. Sci.* **NS-22** (5), 2093-2097 (1975).

1976

92. V.V. Anashin, G.I. Budker, A.F. Bulushev, N.S. Dikansky, Yu.A. Korolyov, N.Kh. Kot, V.I. Kudelainen, V.I. Kushnir, A.A. Livshitz, I.N. Meshkov, L.A. Mironenko, V.V. Parkhomchuk, V.E. Peleganchuk, D.V. Pestrikov, V.G. Ponomarenko, A.N. Skrinsky, B.M. Smirnov, B.N. Sukhina, and A.P. Usov, "Proton storage ring NAP-M, Part 1. Magnetic and vacuum system," *Pribory i Tekhnika Eksperimenta*, No. 4, 31-34 (1976) (in Russian).

93. G.I. Budker, "Charged particle accelerators at the Institute of Nuclear Physics of Siberian Division of U.S.S.R. Academy of Sciences (Novosobirsk) for the national economy," in *Reports of the 2nd All-Union Conf. on Application of Charged Particle Accelerators in the*

National Economy, Leningrad, 1975 (NIIEFA–Efremov Institute, Leningrad, 1976), Vol. I, pp. 48-75 (in Russian).

94. G.I. Budker, Ya.S. Derbenev, N.S. Dikansky, V.I. Kudelainen, I.N. Meshkov, V.V. Parkhomchuk, D.V. Pestrikov, A.N. Skrinsky, and B.N. Sukhina, "Experiments on cooling by electrons, "*Atomnaya Energiya* **40** (1), 49-52 (1976) [*Soviet Atomic Energy* **40** (1), 50-54 (1976)].

95. G.I. Budker, V.A. Gaponov, B.M. Korabel'nikov, G.S. Krainov, S.A. Kuznetsov, N.K. Kuksanov, V.I. Kondratyev, and R.A. Salimov, "ELV-1 electron accelerator for industrial use," *Atomnaya Energiya* **40** (3), 216-220 (1976) [*Soviet Atomic Energy* **40** (3), 256-260 (1976)].

96. G.I. Budker, N.S. Dikansky, V.I. Kudelainen, I.N. Meshkov, V.V. Parkhomchuk, D.V. Pestrikov, A.N. Skrinsky, and B.N. Sukhina, "Experimental studies of electron cooling," *Particle Accelerators* **7** (4), 197-211 (1976).

1977

97. Yu.I. Bel'chenko, G.I. Budker, G.E. Derevyankin, G.I. Dimov, V.G. Dudnikov, G.V. Roslyakov, V.E. Chupriyanov, and V.G. Shamovsky, "Work on high-current proton beams in Novosibirsk," in *Proc. 10th Intern. Conf. on High-Energy Charged Particle Accelerators, Protvino, 1977* (Serpukhov, 1977), Vol. I, pp. 287-294 (in Russian).

98. G.I. Budker, A.F. Bulushev, Ya.S. Derbenev, N.S. Dikansky, V.I. Kononov, V.I. Kudelainen, I.N. Meshkov, V.V. Parkhomchuk, D.V. Pestrikov, A.N. Skrinsky, and B.N. Sukhina, "The status of work on electron cooling," in *Proc. 10th Intern. Conf. on High-Energy Charged Particle Accelerators, Protvino, 1977* (Serpukhov, 1977), Vol. I, pp. 498-509 (in Russian).

99. B.F. Bayanov, G.I. Budker, G.S. Villevald, T.A. Vsevolozhskaya, L.L. Danilov, V.N. Karasiuk, G.I. Sil'vestrov, V.A. Tayursky, and A.D. Chernyakin, "High focal ratio optics with strong magnetic fields for efficiently obtaining secondary particle beams," in *Proc. 10th Intern. Conf. on High-Energy Charged Particle Accelerators, Protvino, 1977* (Serpukhov, 1977), Vol. II, pp. 103-109 (in Russian).

100. G.I. Budker, N.S. Dikansky, I.N. Meshkov, V.V. Parkhomchuk, D.V. Pestrikov, S.G. Popov, and A.N. Skrinsky, "Possibilities for spectrometry experiments with superthin internal targets in charged heavy particle storage rings with electron cooling," in *Proc. 10th Intern. Conf. on High-Energy Charged Particle Accelerators, Protvino, 1977* (Serpukhov, 1977), Vol. II, pp. 141-147 (in Russian).

101. G.I. Budker, A.V. Bulushev, N.S. Dikansky, V.I. Kononov, V.I. Kudelainen, I.N. Meshkov, V.V. Parkhomchuk, D.V. Pestrikov, A.N. Skrinsky, and B.N. Sukhina, "New results from electron cooling studies," in *Proc. 5th All-Union Conf. on Charged Particle Accelerators, Dubna, 1976* (Nauka, Moscow, 1977), Vol. I, pp. 236-240 (in Russian).

102. G.I. Budker, I.B. Vasserman, V.G. Veshcherevich, B.I. Grishanov, M.M. Karliner, V.F. Klyuev, E.V. Kozyrev, I.G. Makarov, O.A. Nezhevenko, G.N. Ostreiko, B.Z. Persov, V.M. Radchenko, R.A. Salimov, B.M. Fomel', Yu.Ya. Chibukov, I.A. Shekhtman, and G.I. Yasnov, "A positron source for the VEPP-4 storage ring," in *Proc. 5th All-Union Conf. on Charged Particle Accelerators, Dubna, 1976* (Nauka, Moscow, 1977), Vol. I, pp. 280-283 (in Russian).

103. V.V. Anashin, V.L. Auslender, G.A. Blinov, L.N. Bondarenko, G.I. Budker, I.B. Vasserman, V.G. Vescherevich, T.A. Vsevolozhskaya, B.I. Grishanov, L.L. Danilov, Ya.S. Derbenev, N.S. Dikansky, B.G. Yerozolimsky, E.I. Zinin, M.M. Karliner, A.V. Kiselev, I.A. Koop, N.A. Kuznetsov, G.N. Kulipanov, B.L. Lazarenko, B.V. Levichev, A.A. Livshitz, A.P. Lysenko, Yu.G. Matveev, N.A. Mezentsev, S.I. Mishnev, A.A. Naumov, V.I. Nifontov, V.N. Pakin, V.S. Panasyuk, Ye.A. Perevedentsev, V.V. Petrov, S.G. Popov, V.P. Prikhod'ko, I.Ya. Protopopov, V.A. Sidorov, G.I. Sil'vestrov, A.N. Skrinsky, E.M. Trakhtenberg, G.M. Tumaikin, V.F. Turkin, B.V. Chirikov, Yu.M. Shatunov, I.A. Shekhtman, L.I. Yudin, and G.I. Yasnov, "Work on colliding electron-positron and electron-electron beams (a review)," in *Fundamental Research (Physical and Mathematical Science)* (Novosibirsk, 1977), pp. 98-104 (in Russian).

104. G.I. Budker, N.S. Dikansky, Ya.S. Derbenev, I.N. Meshkov, V.V. Parkhomchuk, R.A. Salimov, D.V. Pestrikov, A.N. Skrinsky, and

B.N. Sukhina, "The method of electron cooling in charged particle accelerators and storage rings (a review)," in *Fundamental Research (Physical and Mathematical Science)* (Novosibirsk, 1977), pp. 104-107 (in Russian).

105. E.L. Boyarintsev, B.N. Breizman, G.I. Budker, G.E. Vekshtein, V.S. Koidan, V.A. Kornilov, E.P. Kruglyakov, V.M. Lagunov, V.N. Lukyanov, V.V. Mirnov, A.A. Podyminogin, D.D. Ryutov, V.M. Fyodorov, P.Z. Chebotaev, and E.V. Shun'ko, "Investigation of new methods for confinement and heating of a dense thermonuclear plasma (a review)," in *Fundamental Research (Physical and Mathematical Science)* (Novosibirsk, 1977), pp. 114-118 (in Russian).

106. V.V. Anashin, S.E. Baru, G.I. Budker, V.A. Gusev, A.M. Kondratenko, G.N. Kulipanov, E.L. Nekhanevich, N.A. Mezentsev, S.I. Mishnev, G.I. Proviz, G.A. Savinov, V.A. Sidorov, A.N. Skrinsky, E.M. Trakhtenberg, G.M. Tumaikin, A.G. Khabakhpashev, V.B. Khlestov, I.G. Feldman, Yu.M. Shatunov, I.A. Sheromov, and V.N. Shuvalov, "Work on applications of synchrotron radiation (a review)," in *Fundamental Research (Physical and Mathematical Science)* (Novosibirsk, 1977), pp. 119-123 (in Russian).

107. G.I. Budker, V.A. Gaponov, E.I. Gorniker, M.M. Karliner, N.A. Kuznetsov, I.G. Makarov, S.N. Morozov, O.A. Nezhevenko, V.E. Nekhaev, V.S. Nikolaev, G.N. Ostreiko, A.M. Rezakov, R.A. Salimov, G.V. Serdobintsev, A.F. Serov, I.A. Shekhtman, and B.S. Estrin, "High-frequency system of the VEPP-4 electron-positron storage ring, based on the gyrocon — a powerful UHF generator with an unbunched relativistic beam," in *Proc. 5th All-Union Conf. on Charged Particle Accelerators, Dubna, 1976* (Nauka, Moscow, 1977), Vol. I, pp. 284-287 (in Russian).

108. B.F. Bayanov, G.I. Budker, G.S. Villevald, T.A. Vsevolozhskaya, V.N. Karasiuk, G.I. Sil'vestrov, and A.N. Skrinsky, "A system for efficient proton-antiproton conversion," in *Proc. 5th All-Union Conf. on Charged Particle Accelerators, Dubna, 1976* (Nauka, Moscow, 1977), Vol. II, pp. 101-105 (in Russian).

109. G.I. Budker, T.A. Vsevolozhskaya, N.S. Dikansky, I.N. Meshkov, V.V. Parkhomchuk, G.I. Sil'vestrov, and A.N. Skrinsky, "A proposal for proton-antiproton colliding beams based on the electron cooling method for the design of Big Serpukhov accelerator

storage-ring facility," in *Proc. 5th All-Union Conf. on Charged Particle Accelerators, Dubna, 1976* (Nauka, Moscow, 1977), Vol. II, pp. 299-305 (in Russian).

110. G.I. Budker, G.I. Dimov, G.V. Roslyakov, V.E. Chupriyanov, and V.G. Shamovsky, "Storing protons in a ring with the current exceeding the space charge limit," Preprint No. 77-51 of the Institute of Nuclear Physics of the Siberian Division of U.S.S.R. Academy of Science, Novosibirsk, 1977 (in Russian).

111. G.I. Budker, G.S. Villeval'd, V.N. Karasyuk, and G.I. Sil'vestrov, " 'Explosion' parabolic lenses with magnetic fields 0.3-1 MOe," *Zhurnal Tekhnicheskoi Fiziki* **47** (9), 1909-1915 (1977) [*Sov. Phys.-Tech. Phys.* **22** (9), 1107-1110 (1977)].

112. G.I. Budker, "Nuclear physics," in *Za Nauku v Sibiri* (Pro Siberian Science) newspaper, June 10, 1977 (in Russian).

113. G.I. Budker, Ya.S. Derbenev, N.S. Dikansky, I.N. Meshkov, V.V. Parkhomchuk, D.V. Pestrikov, R.A. Salimov, A.N. Skrinsky, and B.N. Sukhina, "Electron cooling and new possibilities in elementary particle physics VAPP-NAP group," in *Proc. XVIII Intern. Conf. on High-Energy Physics, Tbilisi, 1976* (Dubna, 1977), Vol. II, pp. 86-90.

1978

114. G.I. Budker, M.M. Karliner, I.G. Makarov, S.N. Morozov, O.A. Nezhevenko, G.N. Ostreiko, and I.A. Shekhtman, "The gyrocon — a highly efficient converter of energy from powerful relativistic beam for microwave supplies of charged particle accelerators," *Atomnaya Energiya* **44** (5), 397-403 (1978) [*Soviet Atomic Energy* **44** (5), 459-465 (1978)].

115. G.I. Budker, "Electron cooling," *Priroda* (Nature), No. 5, pp. 3-14 (1978) (in Russian).

116. G.I. Budker and A.N. Skrinsky, "Electron cooling and new possibilities in elementary particle physics," *Uspekhi Fizicheskikh Nauk* **124** (4), 561 (1978) [*Soviet Physics–Uspekhi* **21** (4), 277-296 (1978)].

117. G.I. Budker, G.N. Kulipanov, S.I. Mishnev, I.Ya. Protopopov, A.N. Skrinsky, G.M. Tumaikin, and Yu.M. Shatunov, "Storage rings VEPP-2M, VEPP-3, and VEPP-4 as the source of synchrotron radiation," *Nucl. Instrum. Methods* **152** (1), 8 (1978).

1979

118. V.E. Balakin, G.I. Budker, and A.N. Skrinsky, "On the possibility of creating a device with colliding electron-positron beams for ultrahigh energies," in *Proc. 6th All-Union Conf. on Charged Particle Accelerators, Dubna, 1978* (Dubna, 1979), pp. 27-34 (in Russian). Also published in *Problems of High-Energy Physics and Controlled Thermonuclear Fusion (Proc. All-Union Seminar Dedicated to the 60th Anniversary of the Birthday of Academician G.I. Budker)* (Nauka, Moscow, 1981), pp. 11-21 (in Russian).

119. G.I. Budker, M.M. Karliner, I.G. Makarov, S.N. Morozov, O.A. Nezhevenko, and I.A. Shekhtman, "The gyrocon — an efficient relativistic high-power UHF generator," *Particle Accelerators* **10** (1), 41-59 (1979).

1982

120. G.I. Budker, *Collected Works*, edited by A.N. Skrinsky (Nauka, Moscow, 1982). [Items 2, 4-6, 8-14, 19-21, 25, 26, 29, 32, 34, 40, 42-49, 51, 52, 57, 59-62, 64, 65, 67-74, 76, 77, 82, 85-89, 93, 95, 96, 99-102, 107-111, 114, 116-118, and 121-130 of the present bibliography are included in this collection.]

121. G.I. Budker and A.B. Migdal, "Theory of finite lattice" (work completed in 1947), first published in G.I. Budker, *Collected Works* (Nauka, Moscow, 1982), pp. 12-36 (in Russian).

122. G.I. Budker, "On the question of the use of highly absorbing gases for the control and emergency protection of reactors" (work completed in 1949), first published in G.I. Budker, *Collected Works* (Nauka, Moscow, 1982), pp. 36-41 (in Russian).

123. G.I. Budker, "The decrease of power in an atomic reactor when control rods move" (work completed in 1950), first published in G.I. Budker, *Collected Works* (Nauka, Moscow, 1982), pp. 42-47 (in Russian).

124. G.I. Budker and A.B. Migdal, "Deuteron decay due to nuclear forces" (work completed in 1950), first published in G.I. Budker, *Collected Works* (Nauka, Moscow, 1982), pp. 47-49 (in Russian).

125. G.I. Budker and A.A. Naumov, "On the question of the nature and scope of the research work necessary to create a stabilized

electron beam" (work completed in 1953), first published in G.I. Budker, *Collected Works* (Nauka, Moscow, 1982), pp. 150-158 (in Russian).

126. G.I. Budker and V.S. Fursov, "Parametrical resonance in a synchrotron" (work completed in 1947), first published in G.I. Budker, *Collected Works* (Nauka, Moscow, 1982), pp. 230-234 (in Russian).

127. G.I. Budker and V.P. Dmitrievsky, "The effect of n-value azimuthal inhomogeneity on vertical and radial oscillations" (work completed in 1950), first published in G.I. Budker, *Collected Works* (Nauka, Moscow, 1982), pp. 235-240 (in Russian).

128. G.I. Budker, "Calculation of shimming parameters for the correction of local inhomogeneities in the magnetic field" (work completed in 1951), first published in G.I. Budker, *Collected Works* (Nauka, Moscow, 1982), pp. 241-249 (in Russian).

129. G.I. Budker, "On a new method for extracting particles from an accelerator" (work completed in 1951), first published in G.I. Budker, *Collected Works* (Nauka, Moscow, 1982), pp. 249-259 (in Russian).

130. G.I. Budker, N.F. Goncharov, V.A. Gorbunov, R.A. Salimov, I.L. Chertok, and Yu.Ya. Chibukov, "Experiments on extracting an intense accelerated electron beam into the atmosphere" (work completed in 1973), first published in G.I. Budker, *Collected Works* (Nauka, Moscow, 1982), pp. 544-553 (in Russian).

Index